Processes Involved in the Generation of Discolored Water

The mission of the Awwa Research Foundation (AwwaRF) is to advance the science of water to improve the quality of life. Funded primarily through annual subscription payments from over 1,000 utilities, consulting firms, and manufacturers in North America and abroad, AwwaRF sponsors research on all aspects of drinking water, including supply and resources, treatment, monitoring and analysis, distribution, management, and health effects.

From its headquarters in Denver, Colorado, the AwwaRF staff directs and supports the efforts of over 700 volunteers, who are the heart of the research program. These volunteers, serving on various boards and committees, use their expertise to select and monitor research studies to benefit the entire drinking water community.

Research findings are disseminated through a number of technology transfer activities, including research reports, conferences, videotape summaries, and periodicals.

Processes Involved in the Generation of Discolored Water

Prepared by:
P.G.G. Slaats, L.P.M. Rosenthal, W.G. Siegers, M. van den Boomen, R.H.S. Beuken, and **J.H.G. Vreeburg**
Kiwa Water Research

Jointly ponsored by:
Awwa Research Foundation
6666 West Quincy Avenue, Denver, CO 80235-3098
and
Kiwa Water Research
Groningenhaven 7, Postbus 1072, 3430 BB Nieuwegein, The Netherlands

Published by:

Awwa Research Foundation
Advancing the Science of Water®

IWA Publishing

DISCLAIMER

This study was funded by the Awwa Research Foundation (AwwaRF). AwwaRF assumes no responsibility for the content of the research study reported in this publication or for the opinions or statements of fact expressed in the report. The mention of trade names for commercial products does not represent or imply the approval or endorsement of AwwaRF. This report is presented solely for informational purposes.

Published by IWA Publishing, Alliance House, 12 Caxton Street, London SW1H 0QS, UK

Telephone: +44 (0) 20 7654 5500; Fax: +44 (0) 20 7654 5555; Email: publications@iwap.co.uk;
Web: **www.iwapublishing.com**

AwwaRF report number 90966F. Originally published by AwwaRF for its subscribers in 2003.
IWA Publishing version published 2004.

ISBN 1 84339 872 9

CONTENTS

TABLES

FIGURES

FOREWORD

The Awwa Research Foundation is a nonprofit corporation that is dedicated to the implementation of a research effort to help utilities respond to regulatory requirements and traditional high-priority concerns of the industry. The research agenda is developed through a process of grass-roots consultation with subscribers and drinking water professionals. Under the umbrella of a Strategic Research Plan, the Research Advisory Council prioritizes the suggested project based upon current and future needs, applicability, and past work; the recommendations are forwarded to the Board of Trustees for final selection. The foundation also sponsors research projects through an unsolicited proposal process; the Collaborative Research, Research Applications, and Tailored Collaboration programs; and various efforts with organizations such as the U.S. Environmental Protection Agency, the U.S. Bureau of Reclamation, and the Association of California Water Agencies.

This publication is a result of one of these sponsored studies, and it is hoped that its findings will be applied in communities throughout the world. The following report serves not only as a means of communicating the results of the water industry's centralized research program but also as a tool to enlist the further support of the nonmember utilities and individuals.

Projects are managed closely from their inception to the final report by the foundation's staff and large cadre of volunteers who willingly contribute their time and expertise. The foundation serves a planning and management function and awards contracts to other institutions such as water utilities, universities, and engineering firms. The funding for this research effort comes primarily from the Subscription Program, through which water utilities subscribe to the research program and make an annual payment proportionate to the volume of water they deliver and consultants subscribe based on their annual billings. The program offers a cost-effective and fair method for funding research in the public interest.

A broad spectrum of water supply issues is addressed by the foundation's research agenda: resources, treatment and operations, distribution and storage, water quality and analysis, toxicology, economics, and management. The ultimate purpose of the coordinated effort is to assist water suppliers to provide the highest possible quality of water economically and reliably. The true benefits are realized when the results are implemented at the utility level. The foundation's trustees are pleased to offer this publication as a contribution toward that end.

Edmund G. Archuleta, P.E.
Chair, Board of Trustees
Awwa Research Foundation

James F. Manwaring, P.E.
Executive Director
Awwa Research Foundation

ACKNOWLEDGMENTS

The members of the project team are grateful for the financial support provided by the Awwa Research Foundation. In particular, we acknowledge the efforts of our project manager, Kathryn Martin, and our former project manager, Michael Caprara. As well, we acknowledge the insightful contributions made by our Project Advisory Committee. The authors thank them for their time and efforts:

- David Hartman, Cincinnati Water Works, Ohio, United States
- Joop Harms, Vitens, The Netherlands
- Jonathan Clement, Black & Veach, Overland Park, Kan., United States
- David Dixon, The University of Melbourne, Australia

Several utilities in The Netherlands and the Cincinnati Water Works in the United States offered their time, staff, expertise, and facilities for gathering information from the field. These water companies are

- Brabant Water, Den Bosch
- Hydron Midden–Nederland, Utrecht
- Hydron Zuid–Holland, Gouda
- Provincial Water Company Noord–Holland, Velser Broek
- Vitens, Apeldoorn
- Water Company Drenthe, Assen
- Water Company Limburg, Maastricht
- Cincinnati Water Works, Cincinnati, Ohio

EXECUTIVE SUMMARY

BACKGROUND AND APPROACH

The quality of water changes as it passes through the water distribution system. These water quality changes may lead to discolored water at the consumer's tap, which is a major cause of customer dissatisfaction. Controlling water quality changes is of paramount importance for a water company, and control starts with understanding the processes that govern these changes.

The Awwa Research Foundation (AwwaRF) and Kiwa Water Research (Kiwa) have initiated a project to characterize the processes that govern the generation of discolored water and to disseminate knowledge obtained within the joint research program of The Netherlands water companies (BTO-Research). Kiwa has carried out the project within the framework of the AwwaRF–Kiwa partnership.

The project started with a survey that demonstrated that most water utilities must deal with discolored water problems. In general, discoloration is the result of the presence of dissolved, colloidal, or suspended substances in drinking water.

This project focused on the following:

- Source and nature of sediment and corrosion products
- Conditions for resuspension of sediment (hydraulics)
- Formulation and monitoring of water quality conditions for formation of sediment and corrosion products
- Techniques for cleaning water mains

The research on this topic is concentrated in the United Kingdom, the United States, and The Netherlands. Despite the large amount of research on several aspects of discolored water, it is still one of the main concerns of water utilities.

FINDINGS AND CONCLUSIONS

Chemical, physical, biological, and hydraulic processes—both in treatment and distribution—all influence the generation of discolored water. Dominant processes can vary from situation to situation. Almost any change in treatment or distribution will have an effect on the risk of discolored water. The effect may be gradual or momentary.

Processes involved in the generation of discolored water are almost always related. Discolored water can occur only if a source of particulate matter is present and if the hydraulic conditions promote sedimentation and occasional resuspension.

Generation of Discolored Water

Discolored water incidents are the result of sediment resuspension in the network. In general, sediment in the network appears as

- Suspended solids in flowing water
- Attachments to the pipe wall (for instance, as biofilm or a corrosion layer)
- Loose deposits at the bottom of the pipe

The most relevant sources for sediment in the distribution system are identified as

- Directly from the treatment plant. Treatment plant operation can result in particulate matter getting into the distribution system. Almost all treatment processes based on filtration introduce sediment in the network. Iron plays an essential part in this.
- Corrosion of water mains or domestic installations. Corrosion of unlined cast-iron water mains seems to be one of the most dominant processes here. The mechanisms of iron release into bulk water to form discolored water are still not fully understood.
- Installation and maintenance work in the network. The distribution system may be fouled as a result of less careful installation or maintenance work, which allows sand, clay, and other particulate matter to enter the system.
- Formation within the distribution system by precipitation of dissolved elements (such as iron, manganese, calcium, and silicates) or sediment formation. This is caused by the growth of micro- and macroorganisms in the water mains.

Hydraulic processes largely determine whether sediment will lead to discoloration or not. They are a key factor to consider when dealing with discolored water problems.

Discoloration is the result of a combination of factors; for instance, the presence of sediment combined with hydraulic disturbance or active corrosion in combination with large retention time. Consequently, a multiple-stage approach is necessary for controlling discoloration.

Techniques for Discolored Water Risk Assessment

The occurrence of discolored water can be monitored on the basis of

- Customer complaints
- Sediment and turbidity monitoring (various sampling techniques)
- Resuspension potential (Kiwa Water Research, The Netherlands)
- Filter paper color method (WRc, United Kingdom)

However, details of each method's relationship to customer perception thresholds are not available.

Two successful techniques for risk assessment were further developed:

1. Complaint registration about discolored water. This is a useful method when a protocol for interpretation is available, such as the classification system developed in the United States. In such a system, complaints are combined with other criteria leading to concrete conclusions.

Table ES.1
Criteria for degree of fouling converted from the Kiwa method

Pipe fouling	Kiwa1 (SIGRIST* tap-in point, ftu)	Kiwa2 (SIGRIST fire hydrant, ftu)	OPM† (Dr. Lange fire hydrant, ftu)‡	PWN§ (random samples, ftu)
Clean	≤1	≤1.5	≤8	≤13
Slightly fouled	>1 and ≤2	>1.5 and ≤3	>8 and ≤25	>13 and ≤28
Fouled	>2	>3	>25	>28

*Turbidity meter from SIGRIST-Photometer AG, Ennetbürgen, Switzerland.
†OPM = simplified resuspension potential method.
‡Turbidity meter from Dr. Bruno Lange GmbH & CO. KG, Düsseldorf, Germany.
§PWN = Provincial Water Company of North Holland.

2. Actual measuring of the mobility of sediment and the risk for discoloration. The resuspension potential method is probably the most useful because it characterizes sediment based on resuspension behavior. This technique is, however, rather complex and time consuming (and thus expensive). Much research has been conducted into the method's optimization and simplification. The method can currently be carried out relatively easily in a short period of time. The results of a method largely depend on the way the method is implemented and on the equipment that is used.

Table ES.1 shows the criteria used in The Netherlands for the resuspension potential that indicate that a water main system is fouled. However, interpretation of results requires skill and experience.

Discolored Water Risk Management

The combination of complaint registration and water quality monitoring allows the risk for discolored water to be determined. Derived from the risk assessment and the understanding of how discolored water is generated, actions are formulated to prevent discolored water problems and manage the risk. Three steps are distinguished:

Step 1: Prevent particles from entering the network
Step 2: Prevent suspended solids from settling
Step 3: Remove sediment from mains effectively and in due time

Each step can be implemented separately, fighting one of the causes of discoloration. The combination of steps will give the best result.

Step 1: Prevent Particles From Entering the Network

Water quality at the treatment plant may be improved in terms of

- Iron and manganese removal
- Biostability of the drinking water
- Corrosivity for cast iron

Most processes that influence water quality also influence each other. Biological processes influence sediment characteristics; corrosion products influence biological processes; sediment may influence corrosion processes, and so on. Therefore, an integral strategy is needed to simultaneously deal with as many quality-threatening processes as possible. A cost-benefit analysis should determine which control measures ought to be applied in which circumstances.

The quality of operations and maintenance, and the procedures used for these tasks, are of paramount importance for controlling water quality in both production and distribution.

Treating the drinking water can restrict the corrosion of cast iron. The pH and the total inorganic carbon content of the water need special attention. Increasing the buffer capacity decreases iron release. The effectiveness of adding phosphates to combat discolored water is still under discussion. Silicate addition decreases iron release by cast iron.

The filtrate scan appears to be a useful instrument for monitoring and quantifying the effect of optimizing treatment to reduce particles in the finished water. Turbidity and particle counting are especially useful monitoring techniques.

Step 2: Prevent Suspended Solids From Settling

Sediment behavior in distribution pipes has been the subject of research conducted in The Netherlands in recent years. The starting point was to emulate sediment behavior for a limited number of sediment types in a pipe with a smooth wall, without biofilm, for a short measuring period. This showed that the velocity of water in which the first sediment particles start to move is approximately 0.10 m/sec for light sediment and approximately 0.15 m/sec for heavier sediment. The light particles are all in motion at a velocity of about 0.15 m/sec, the heavier ones at a velocity of approximately 0.25 m/sec.

No relationship was observed between the pipe diameter and the velocity at which all particles are in motion. However, it became clear that, for an equal velocity, the velocity at which sediment moves through a pipe increases as the diameter decreases.

At the velocities that occur in the water main system during normal use, most sediment with a diameter of less than 50 μm will roll along the bottom of the pipe. Lighter material that is resuspended will mainly stay in the lower half of the pipe.

For self-cleaning distribution networks, a design velocity of 0.4 m/sec is used in combination with the current peak consumption method (the $q\sqrt{n}$ method, which is a calculation method for estimating the total flow [q] at a number [n] of taps). However, there are indications that the $q\sqrt{n}$ method overestimates the actual peak consumption that occurs.

Step 3: Remove Sediment From Mains Effectively and in Due Time

The third step consists of removing the sediment from the water main system. Water flushing is the most effective technique. When flushing, the flushing rate of the water must be at least 1.5 m/sec. The volume of flushing water must be at least 3 times the volume of water in the pipe, until a visual inspection shows that the water is clean. The work is performed on the basis of a clean-water front. Making measurements using the resuspension potential method, in addition to carrying out the cleaning activities, increases effectiveness.

In practice, water flushing has been found to involve a number of limitations, such as the flushing of cast-iron pipes or restrictions on the supply capacity of pipes or on the discharging of the flushing water. In these cases, newly developed techniques can be used, such as pulsating flushing, ball-enhanced flushing, or the FLOWJET (WRc, United Kingdom) technique.

With pulsating flushing, accelerations during flushing are introduced in a relatively simple and controllable manner. Tests show that a clearly observable cleaning effect occurs with pulsating flushing.

Ball-enhanced flushing is likewise an effective method of removing sediment from pipes. This technique is mainly suitable for pipes with a diameter of more than 250 mm and in which restrictions exist on the volume of water that must be discharged. For the time being, pulsating flushing appears to be a better technique for remote pipes with insufficient supply capacity.

The systematic approach to using flushing to clean water main systems is relatively expensive. It is, therefore, important to clean the water main system at the right moment—when a pipe is fouled to just below the level that would cause customers to complain. The moment can be estimated by making monitoring measurements. The extra costs of measuring should be in balance with the savings that can be achieved by deferring cleaning.

A measuring density of one measurement per 10–20 km has been estimated to result in an acceptable cost level. The savings achieved by making a realistic reduction in the cleaning frequency appears to be larger than the extra costs of making the measurements. If a deeper insight is obtained about the degree of fouling in distribution networks after a number of years, the measuring density can be adjusted.

CHAPTER 1
INTRODUCTION

Every drinking water network throughout the world must occasionally deal with complaints about discolored water, and it is one of the main reasons why customers complain to drinking water suppliers about water quality. In general, discoloration is the result of the presence of dissolved, colloidal, or suspended substances in drinking water. These substances may originate from the treatment plant, be the products of corrosion processes, or form in the network. They can enter drinking water directly or after precipitation followed by resuspension in changing flow conditions.

Periodic cleaning programs are aimed at curing or preventing customer complaints. To justify the energy that is put into these cleaning programs, water companies must ensure that the programs have the desired effect. Several cleaning methods are used, such as flushing, water–air flushing, and pigging, but conclusions on the effectiveness are unclear. Apart from that, adequate cleaning programs take care of the problem only for a time. To be able to find more lasting solutions, water utilities must fully understand the mechanisms that cause discolored water.

The Awwa Research Foundation (AwwaRF) and Kiwa Water Research (Kiwa) have undertaken a project to characterize the processes that determine the generation of discolored water. Kiwa has carried out the project within the framework of the AwwaRF–Kiwa partnership. This research program focused on three questions:

1. What is the cause of the discolored water? This is the central question because answering it is key to solving the problem. Most customers complain about "colored" water using terms such as brown, red, black, white, or even blue water.
2. How can we measure water discoloration? Discolored water in a network is almost always a temporary problem. The symptoms may be gone completely only a few minutes after a customer has complained.
3. What can be done about it? Because sediment is the main cause of discolored water, several processes may result in discolored water at the consumer's tap. The processes taking place in the network form the basis for deciding how to prevent discolored water from occurring.

This report presents the results of research conducted to characterize the processes that are involved in the generation of discolored water and discusses what can be done about it. Chapter 2 describes the types of discolored water, and chapter 3 describes the processes involved in discolored water generation. Chapter 4 provides an overview and details of developments in the methods of detecting discolored water, and chapter 5 outlines a three-step approach to discolored water risk management. Chapters 6, 7, and 8 describe the three steps for managing the risk of discolored water:

- Step 1: Prevent particles from entering the network
- Step 2: Prevent suspended solids from settling
- Step 3: Remove sediment from mains

Finally, chapter 9 presents the conclusions and recommendations.

CHAPTER 2
CHARACTERIZATION OF DISCOLORED WATER

DEFINITIONS OF DISCOLORED WATER

Complaints about discolored water are one of the most frequent causes of consumer dissatisfaction with water utilities (Vreeburg 2000). Most customers complain about colored water using terms such as brown, red, black, white, or blue. However, the literature offers no general definition of discolored water.

In a recent AwwaRF study on the development of red-water control strategies, red water was defined as an aesthetic property of the water, caused in part by the presence of iron, that results in discolored water (red, brown, or yellow) at the consumer's tap.

Discolored water may originate from both ground- and surface waters. Sources of iron may be from naturally occurring iron deposits that are taken up by groundwater while the water is still in the ground, from the iron salts used in the coagulation process at the water treatment plant, and from iron-containing materials in the water distribution and piping systems on the premises (Clement et al. 2002).

In addition, there is also no uniform way of describing discolored water. Turbidity, particle size, color measurements, metal concentrations, iron and manganese contents, and even questionnaires are used (Liebeschuetz and Hulsmann 1990). However, none of these determinants expresses the total impact of discolored water. Turbidity is, for example, a collective parameter that is related to all kinds of solids in the drinking water. Turbidity is measured using a turbidity sensor, but sensors vary, and turbidity measurements obtained with one type of apparatus are not the same as those obtained with a sensor of another type (Slaats 2000).

Filtering with varying pore sizes gives some information about the size of the particles present. Color measurements are expressed in milligrams per liter platinum–cobalt (Pt–Co), but color gives no information about the substance that causes the color in the water.

In the United States, the term *apparent color* is used. The term *color* is normally used to indicate *true* color (i.e., the color of water from which turbidity has been removed). Apparent color refers to color resulting both from substances in solution and from suspended matter. Apparent color is obtained without filtering the original sample.

Most cases of discolored water refer to red or brown water, and the literature predominantly focuses on this type of discoloration (Clement et al. 2002). However, it is a misconception that the presence of red or brown water is directly related to corrosion of cast-iron mains. Red or brown water also appears in distribution networks with no cast-iron pipes. Insufficient treatment of the water or the formation of sediment in the network can also contribute to discolored water (Vreeburg 2000).

Red or brown water can also originate from domestic drinking water installations when galvanized steel pipes are used (Rhodes Trussell and Wagner 1996).

Black water, caused by the formation of ferrous iron, can be produced if the water's oxygen content is depleted before red or rusty ferric iron is formed. Black water can also be caused by manganese dioxide formation, resulting from the presence of manganese in the source water (Kirmeyer et al. 2000).

White water may be the result of excessive aeration, and degassing of the water solves this problem. In other cases, white water can be caused by the presence of calcium carbonate particles.

Blue water is a phenomenon that is reported occasionally and generally occurs as a local outbreak (Taylor, O'Halloran, and Cannington 1996). Blue water is caused by the presence of copper corrosion products caused by excessive corrosion of copper pipes. Extensive effort has been put into a better understanding of the factors that cause outbreaks of blue water. However, little is known about factors that initiate or propagate blue water (Edwards, Jacobs, and Taylor 2000).

Yellow water can be caused by too high a content of humic acids or natural organic matter (NOM) in the water. The most widely used method of decreasing the color caused by NOM is to treat the water with activated carbon.

In Dutch research, discolored water is defined as "water with a turbidity that is noticed by a customer" (Vreeburg 2000). This definition excludes water with a color that is not caused by discrete particles, such as in white water with an oversaturation of oxygen or blue water caused by copper corrosion products. Based on this definition, the current research on discolored water focuses on the sediment in the distribution systems.

NATURE OF SEDIMENT

Discolored water is the result of sediment resuspension in the network. In general, sediment in the network is found

- As suspended solids in flowing water
- Attached to the pipe wall; for example, as biofilm or a corrosion layer
- As loose deposits at the bottom of the pipe

A great deal of effort is put into analyzing the composition of sediment in the network. Smith and colleagues have studied deposits in water mains by collecting samples both by hydrant flushing and by removing sections from steel pipes (Smith et al. 1997). She concluded that deposits predominantly consisted of the compounds siderite ($FeCO_3$), magnetite (Fe_3O_4), lepidocrocite (γ-FeOOH), goethite (α-FeOOH), calcite ($CaCO_3$), and quartz (SiO_2). Iron accounted for approximately one quarter of elements present in the sample, and other minor constituents were silicon, sulfur, and calcium.

The nature of sediment in a distribution system is also studied by sampling from reservoirs and pipes (Gauthier et al. 1996). The elementary mineral and organic composition, bacterial counts, volatile fraction, and appearance were analyzed in all the deposits. The most remarkable feature was that the compositions of deposits varied greatly from one sampling point to another within a large distribution system that was supplied from a single source of water. The researchers found that the quantity of iron in the sediment increased during periods of stagnation in the water main. It was notable that, in all pipes and also in samples obtained from flushing polyvinyl chloride (PVC) mains, the deposit samples from the network had high iron content. The researchers suggest that those small particles entering from the treatment plant accumulate to form large particles in the distribution network. During the investigation, the sediment was collected in a 100-µm net. The particles were not detected in the reservoir sediment, but they were detected after they accumulated in the distribution main. The overall conclusion from Gauthier and co-workers was that drinking water systems carry water, minerals, and organic matter. Many of these materials are dissolved, but a small fraction can settle out to form loose sediment in the distribution system.

The differences in composition among sediments in the same network indicate that several mechanisms (physical, chemical, and biological) govern the formation of deposits. All deposits

Table 2.1
Sediment volume of different fractions obtained by flushing on hydrant (flushing velocity, 1 m/sec)

Fraction	Average (mL/m^3)	50th percentile (mL/m^3)	90th percentile (mL/m^3)	Maximum value (mL/m^3)
>500 μm	1.3	0.5	3	100
100–500 μm	33	7	90	1,500
30–100 μm	147	32	321	9,000

Table 2.2
Volume of sediment for different pipe materials (flushing velocity, 1 m/sec)

Pipe material and distribution of sampling points	Asbestos cement (54%)	Cast iron (8%)	PVC (35%)	Other (3%)
Particles >500 μm (mL/m^3)	1	6	1	2
Heavy particles 100–500 μm (mL/m^3)	1	3	1	0
Light particles 30–100 μm (mL/m^3)	30	50	40	30

were heavily colonized with microorganisms, indicating that sediment provides a porous medium that allows bacteria to attach and grow, especially when the aluminum fraction is high. The sediment is also a source of organic nutrients for microorganisms and thus can be considered as a biomass reservoir within the distribution system. To prevent occurrences of discolored water, it is important to prevent the accumulation of sediment in drinking water systems.

In The Netherlands, the composition of sediment and the presence of invertebrates have been analyzed in sediment samples obtained via hydrant flushing and division by sieving the sediment into portions of equal size (van Lieverloo et al. 2002). The sediment fractions are divided into three categories: particle sizes between 30 and 100 μm, those between 100 and 500 μm, and those greater than 500 μm. The samples were obtained from several types of water treatment plants and pipe materials. The samples were analyzed using light microscopy and specific element analyzers.

The results showed that the fraction with particle sizes between 30 and 100 μm accounted for the largest volume of the sediment (Table 2.1). The particles that were greater than 500 μm consisted mainly of heavy materials, such as rust particles, sand, and pipe materials. The particles that were between 100 and 500 μm in size consisted mainly of organic matter with a rather low density. The fraction of 30–100 μm can also be expected to consist mainly of organic matter.

Table 2.2 shows the distribution of heavy and light particles in relation to pipe material. We can see that, relatively, most sediment was present in cast-iron pipes and consisted of light material. It should be noted that the sediment samples in this study were obtained by flushing at a velocity of 1 m/sec. This is not necessarily representative of the situation under normal hydraulic conditions.

CHAPTER 3
GENERATION OF DISCOLORED WATER

INTRODUCTION

In general, the processes involved in the generation of discolored water are

- Processes that lead to particles entering the distribution system:
  - Directly from the treatment plant
  - From corrosion of water mains or domestic installations
  - From inadequate installation or maintenance work
- Processes that lead to particle formation within the distribution system by physical, chemical, and biological processes (ongoing coagulation, formation, and abrasion of biofilm)
- Hydraulic processes that lead to accumulation and resuspension of particles

Figure 3.1 shows the model generally used in The Netherlands to yield an overview of these processes (Vreeburg 1996). Discolored water occurs if the outgoing turbidity, C_{out}, becomes too high.

PROCESSES THAT LEAD TO PARTICLES ENTERING THE DISTRIBUTION SYSTEM

Basically, there are three ways for particles to enter the distribution system:

1. Directly from the treatment plant. One of the recognized causes of discolored water is insufficient water treatment at the plant. Poor treatment plant operation can result in particulate matter getting into the distribution system and causing discoloration and other problems (Gray 1994). This is caused mainly by using excessive amounts of coagulants, which results in either iron or aluminum entering the system, although manganese, algae, and organic matter can all arise from poor treatment as well.

   Iron appears to play an important role in this process. It is certainly part of most discolored water problems and can be expected to cause them. Dissolved iron will not be removed in treatment and will not settle. However, it can be oxidized in the presence of strong oxidants (such as chlorine dioxide [ClO_2] and hypochlorous acid [HOCl]) and settle in the network. Generally, iron colloids are not removed by sand filtration and can play an important role in sediment formation.
2. Corrosion of water mains or domestic installations. Corrosion of iron mains is very complex and has been intensively studied since the beginning of the last century.

   Iron release, red water, and the Kuch mechanism have been discussed and reviewed exhaustively (Benjamin, Sontheimer, and Leroy 1996; McNeill and Edwards 2001). Research into iron release from unlined cast-iron pipe was recently published and included descriptions of methods for controlling the red water that forms on the basis of this phenomenon (Clement et al. 2002).

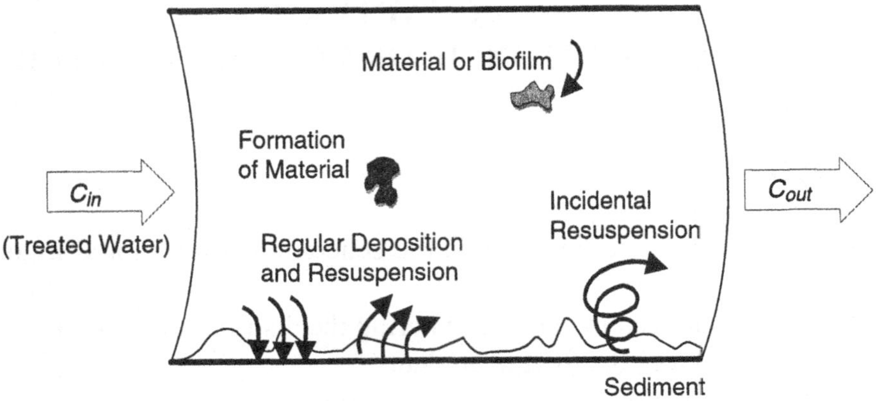

Figure 3.1 Overview of the processes involved in the generation of discolored water

It was not the objective of this project to mention all the results of research conducted in the past. However, we can conclude that corrosion of unlined cast-iron water mains may result in thick encrustation on the inner surfaces of the unlined cast-iron pipes. Depending on the hydraulic conditions, iron scale material may be dislodged by water flushing and then settle in the distribution system as particulate matter. Another iron release mechanism is dissolution, where oxidized iron ions are continuously released into the water over time instead of becoming part of the scale. Clement et al. (2002) concluded that the mechanisms of iron release into bulk water to form discolored water are not fully understood and specifically recommend conducting a project devoted entirely to understanding iron scales.

3. Inadequate installation and maintenance work. The distribution system may be fouled as a result of careless installation or maintenance work that allows sand, clay, and other particulate matter to enter the system.

PROCESSES THAT LEAD TO PARTICLE FORMATION WITHIN THE DISTRIBUTION SYSTEM

It is recognized that precipitation of dissolved elements (such as iron, manganese, calcium, and silicates) or the growth of micro- and macroorganisms in the water mains can form sediment.

As shown by the study into the nature of sediment in a distribution system (Gauthier et al. 1996), drinking water systems carry water, minerals, and organic matter. The differences in composition among sediments in the same network indicate that several mechanisms (physical, chemical, and biological) govern the formation of deposits.

Treated water is supposed to have a high degree of chemical and biological stability so as to prevent undesirable changes in water quality during transportation, storage, and distribution. Degradation of the microbial quality of drinking water in the distribution system can largely be attributed to biofilm formation and sediment accumulation. Growth of microorganisms can lead to consumer complaints about the taste, odor, or discoloration of water (van der Kooij et al. 1999). Therefore, water leaving the treatment plant and materials used in drinking water distribution should be biostable.

Further, inorganic particles such as hydroxides/oxides of iron and manganese can be trapped in biofilms. Sloughing of biofilms can enhance sediment accumulation in peripheral parts of the distribution system. The sediment provides an environment in which certain types of microorganisms (such as *Aeromonas* and invertebrates) can multiply. To understand and prevent these undesirable quality changes, the growth-promoting properties of water and materials in contact with water have been evaluated in The Netherlands (van der Kooij et al. 1999).

HYDRAULIC PROCESSES THAT LEAD TO ACCUMULATION AND RESUSPENSION OF SEDIMENT

Hydraulic processes, although a key factor to consider when dealing with discolored water problems, have received little attention until quite recently.

Consumer complaints about discolored water usually stem from hydraulic events. Changes in the hydraulic regime can cause particles to resuspend, leading to a high-turbidity peak at the consumer's tap.

Typical hydraulic events that lead to discolored water are sudden changes in the direction or velocity of the water that exceed the changes that occur during regular operation. These changes may occur for various reasons; for example, a water main burst, opening or closing valves, bringing mains back into use after repair, and the use of water for fire fighting.

It is also internationally recognized that stagnation time influences the occurrence of discolored water. The most frequently cited reason for discolored water is the presence of dead-end mains. However, dead ends can be seen as an exceptional case of a change in the flow of the water. Because of the absence of flow, particulate matter can easily settle in dead ends; when water flows, this material will be resuspended.

Determining the effects of hydraulic processes on the generation of discolored water is a complex process that requires insight into the hydraulic conditions and the hydraulic characteristics of the sediment. The material that is most likely to cause discoloration is the small and lightweight material, which settles very slowly and resuspends easily when flow changes occur. It follows from Table 2.2 that lightweight sediment is dominant in terms of volume and can be found in all systems, regardless of pipe material.

CONCLUSIONS

Although the processes involved in the generation of discolored water are known in a general sense, the mechanisms that govern these processes are not fully understood. This is especially true for

- The iron release mechanisms
- The formation of particles within the distribution system and the interrelationships among the physical, chemical, and biological processes that determine particle formation
- Sediment behavior as a result of hydraulic conditions and sediment characteristics

It is important to understand that the processes are almost always related. Discolored water can occur only if a source of particulate matter is present and the hydraulic conditions promote sedimentation and occasional resuspension.

CHAPTER 4
TECHNIQUES FOR DISCOLORED WATER RISK ASSESSMENT

INTRODUCTION

Water utilities are not, in general, aware of the fact that they are delivering discolored water. Complaints received from customers, who notice changes in the normal organoleptic appearance (color, turbidity, taste, and odor) of the water, may be the utility's first indication of a problem. In the case of organoleptic deviations of water, samples of tap water are analyzed for chemical determinants such as iron, aluminum, manganese, turbidity, and color. However, analyzing water samples is, in most cases, not a successful way to detect discolored water, as the color has generally already disappeared. In practice, only a few methods have been developed for predicting discolored water at the tap or in the field. These methods are based on gathering knowledge on consumer perceptions (complaints) and on detecting sediment in the drinking water (at the tap or in the field).

CONSUMER COMPLAINTS

Consumer complaints about discolored water provide a general parameter that is used as a basis for taking actions in the distribution system. Registration systems and databases on this subject are sometimes available.

Some water utilities have developed their own approach. Efforts to enhance the quality of water delivered to customers of the Louisville Water Company, Louisville, Ky., with 2,900 mi (4,667 km) of distribution mains, have shifted in emphasis from the original approach of upgrading the treatment process to examining the nature and performance of the distribution system. Complaints about taste and odor problems (and reports of discolored water) from consumers were analyzed to establish priorities for implementing mains rehabilitation and treatment programs. Because no consistent pattern of complaints relating to the entire network could be detected, the reports were analyzed by reference to the several-meter districts. This allowed particular local problems that could receive remedial treatments of a site-specific nature to be identified. Such local problems included the prevalence of dead ends, periods of long standstill, the presence of unlined cast-iron mains, and low flows (because of low population densities and seasonal, or other, off-peak periods of low consumption). By matching the potential risks with the incidence of reported complaints, a system of priorities for corrective actions (such as selective finishing, mains relining, and cross linking) could be devised (Davis and Smith 1993).

The Water Department of the City of Topeka, Kan., has maintained detailed records of consumer complaints since 1988 (Clement et al. 2002). Consumer complaints were used as a measure of treatment effectiveness. Although this parameter does not have the precision typically associated with laboratory analyses, it reflects one of the primary objectives of water treatment. The complaint index was calculated by assigning a value of 100 to the maximum number of calls per month received during the 10-year period covered by the investigation. Graphs were constructed to determine if relationships existed between consumer complaints and water quality parameters. From the data obtained by analyzing complaints in this way, only the relationship between the complaint index and the alkalinity of the water yielded useful information for plant operation. A final water alkalinity level in excess of 60 mg/L appeared to substantially reduce the number of complaints about discolored water.

The Customer Complaint and Red-Water Classification System was developed during a study of strategies for controlling red water (Clement et al. 2002). This system is essentially an organized method of gathering, categorizing, and evaluating data to obtain an understanding of the sources, types, and causes of red water. Information is gathered in four key areas:

1. Customer information
2. Water supply system characteristics
3. Water supply system operations
4. Water quality and scale analysis

With this information, the utility is in a much better position to understand the problem and propose sound and cost-effective solutions. The system will enable the utility to address the immediate problem and then to systematically determine whether the long-term problem is acute or chronic. Acute problems can often be addressed by changing operational procedures, whereas chronic problems often require changes in sources or treatment, or rehabilitation of piping systems that are the underlying cause of the problem. The details of the system were published only recently, before this project was finalized in 2002. No information obtained in daily practice is available yet from the use of the Customer Complaint and Red-Water Classification System.

Customer complaints, an important source of information for the water company, can spur the cleaning of the water main system. As mentioned previously, sediment can build up in the water main system. At high velocities caused by peak demand or flow direction reversal, the sediment in the water main system may be resuspended, which may result in complaints about discolored water at the tap. The number of complaints is expected to increase as the degree of network fouling increases, and the number of complaints per connection can give an indication of overall customer perception. Using complaints about discolored water as an indicator of the degree of network fouling enables the water main system to be managed more efficiently.

This is apparent from the complaint records of the water company NRE-WATER, which supplies the city of Eindhoven in The Netherlands. In the early 1990s, NRE-WATER had a high number of complaints about discolored water. Consequently, the company carried out an extensive cleaning program in 1992 and 1993. The number of complaints per year fell markedly after this, from 633 in 1992 to 104 in 1995.

After 6 years, the number of complaints had risen again (Figure 4.1). In 1996, 145 complaints were recorded, and in 1998 the number had increased to 252 (Schaap and Ekkers 1999).

If a figure of no more than 250–300 complaints per year is used as the quality limit, the time between two cleaning operations is approximately 5 years.

MONITORING TECHNIQUES

Filter Color Method

In the United Kingdom, WRc plc developed the filter paper color (FPC) method to link discolored water to consumer perception. The FPC can be measured manually or by using an automatic liquid filtration instrument (ALFI; Williams 1990). The principle of the method is to grade the color of the stain left on a filter paper after filtering a set volume of test water. The

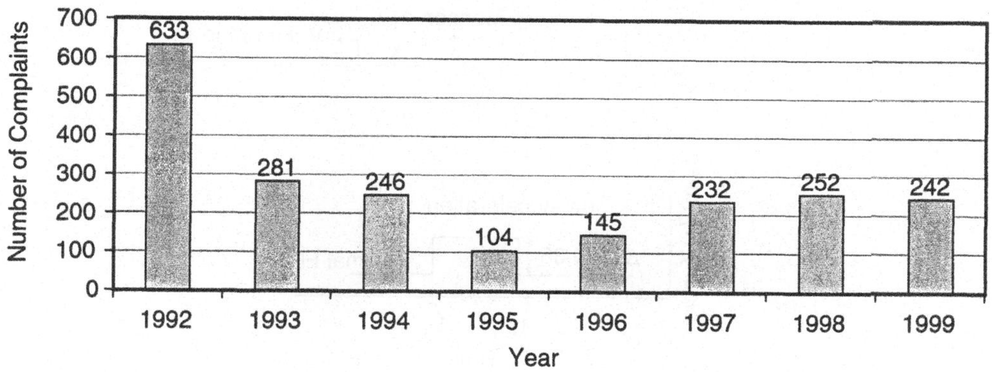

Figure 4.1 Discolored water complaint records of NRE-WATER from 1992–1999

method measures the colored particulate matter and gives a direct indication of discoloration as perceived by the consumer. The method can distinguish between colorless and visibly discolored water. The application of ALFI includes examining the cause of discolored water problems; investigating specific, persistent consumer complaints; and monitoring the performance of treatment works.

Resuspension Potential Method

Because discolored water samples are difficult to obtain, a method was developed in The Netherlands to predict discolored water events in the network, the so-called *resuspension potential method* (Vreeburg 1996). This method is based on the principle that by introducing a disturbance in the water flow, sediment present in the pipeline is resuspended. The disturbance to be introduced is standardized and reproducible. It has been empirically determined that a disturbance giving a flow increase of 0.35 m/sec on top of the normal flow rate will meet these requirements. The disturbance is maintained for 15 minutes. The turbidity of the water in the water main system is continuously measured before, during, and after the disturbance.

The method of resuspension is carried out with monitoring systems that are connected at various specific locations in the water main system. Apart from turbidity, parameters such as pH and conductivity can be measured. The data are stored electronically for further processing and interpretation. Figure 4.2 gives an overview of the principle of the resuspension potential method. The change in turbidity during the disturbance is a measure of the quantity of resuspension of the sediment (Figure 4.3). Currently, many water utilities use this method to systematically monitor the presence of sediment in the network. Until now, no problems have been experienced with implementing the resuspension method in old cast-iron pipes or in asbestos–cement (A–C) pipes.

Simplified Resuspension Potential Methods

Numerous water companies in The Netherlands have given their own shape to the resuspension potential measurement by using various sampling points (such as the service pipe or fire hydrant) and various turbidity meters and measuring methods (online, grab samples). Because these simplified methods take less time, they are cheaper to perform. The following simplified resuspension potential methods are currently available:

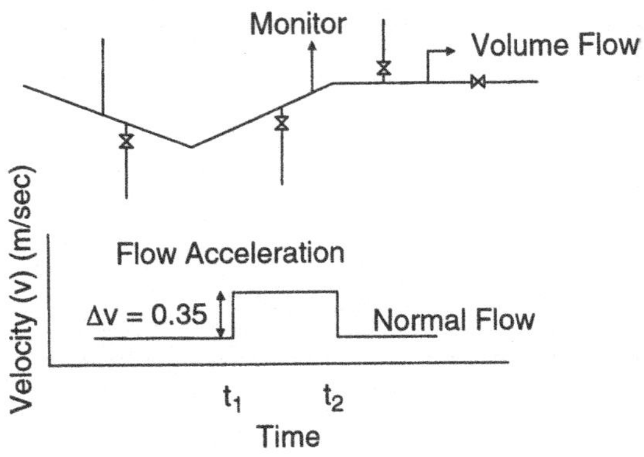

Figure 4.2 The principle of the resuspension potential: a flow of 0.35 m/sec is introduced over a period of 15 minutes while turbidity is continuously measured

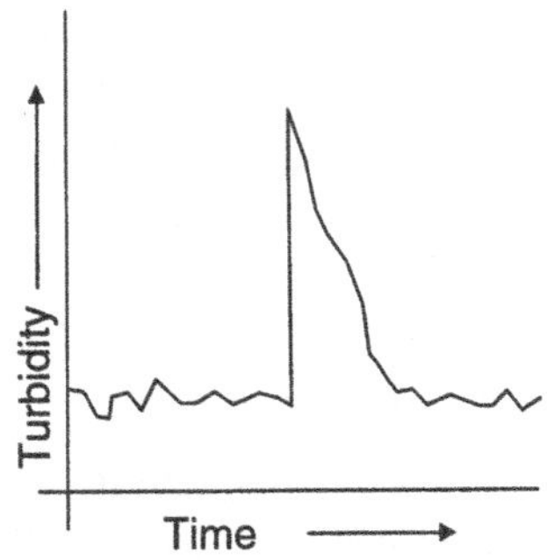

Figure 4.3 Change in turbidity during disturbance of flow velocity

- The OPM method (developed with the Dutch BTO-Research group)
- The PWN method (developed by the Provincial Water Company of North Holland in Velser Broek, The Netherlands)
- The Kiwa method (developed by Kiwa Water Research in Nieuwegein, The Netherlands)

The OPM Method

The OPM method, which is a resuspension potential measurement, involves using a Dr. Lange Ultraturb turbidity meter (Dr. Bruno Lange GmbH & CO. KG, Düsseldorf, Germany) to measure turbidity continuously. The meter is connected to a fire hydrant in the water main system. Figure 4.4 is a diagram that shows the implementation principles of the OPM method, which are

- Disturbance through fire hydrant
- Continuous turbidity measurement in the water from the same fire hydrant

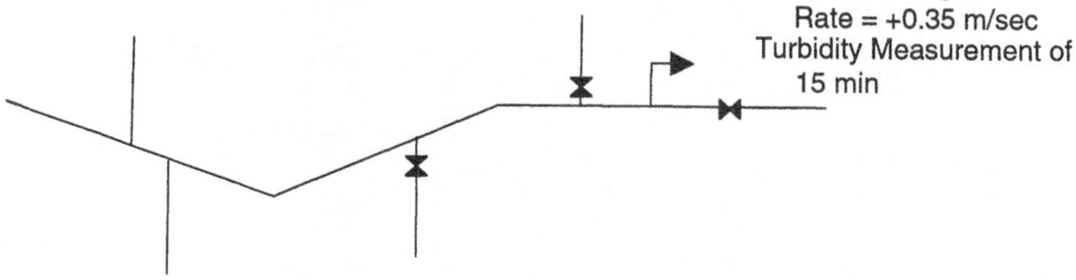

Figure 4.4 The principle of the OPM method

The OPM method consists of

- Establishing a definite flow direction by setting valves 24 hours before the measurement will take place.
- Using a Dr. Lange turbidity meter to measure turbidity continuously through the fire hydrant during the disturbance. The data logger is set to record 1 measurement per minute.
- Measuring (or recording) 10 minutes of turbidity.
- Opening a fire hydrant for 15 minutes so that the velocity increase in the pipe is 0.35 m/sec.
- Measuring (or recording) 15 minutes of turbidity.

The PWN Method

With the PWN method, a disturbance is introduced in a fire hydrant. Grab samples of the water are taken from the fire hydrant during the disturbance. The main principles of the PWN method are

- Disturbance through fire hydrant
- Water samples taken from the same fire hydrant
- Turbidity determined in the laboratory

The disturbance is introduced using a volume flow restriction to create similar velocity changes in pipes of different diameters. Figure 4.5 shows a diagram of the method, which consists of

- Establishing a definite flow direction by setting valves and waiting for about 1 day for a stable situation
- Opening a fire hydrant with a volume flow restriction for approximately 15 minutes

During the period of the disturbance, four 100-mL samples are taken (after 3, 6, 9, and 12 minutes) and the turbidity is determined in the laboratory. The average turbidity of the four samples is adopted as the measure of fouling.

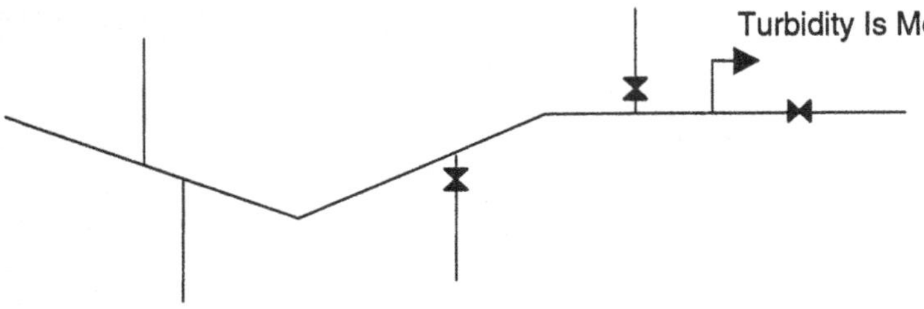

Figure 4.5 The principle of the PWN method

The Kiwa Method

The Kiwa method was developed to determine the fouling mechanism as well as the degree of network fouling. For this reason, the measurement generally takes a few days. A disturbance is created halfway through the measurement.

To compare the various methods, the Kiwa method was shortened to encompass only the duration of the disturbance. Kiwa's monitoring systems are fitted with turbidity meters made by SIGRIST-Photometer AG (Ennetbürgen, Switzerland). The main principles of the Kiwa method are

- Tap-in point in the pipe being investigated
- Disturbance through fire hydrant
- Continuous measurement of turbidity from tap-in point

Figure 4.6 shows how to measure the degree of network fouling. The method consists of

- Establishing a definite flow direction by setting valves and waiting for approximately 1 day for a stable situation
- Increasing the velocity in the pipe being investigated by 0.35 m/sec for 15 minutes by adjusting a fire hydrant
- Measuring through a tap-in point

As shown in Figure 4.3, the size of the turbidity peak during the disturbance (the resuspension potential) and the duration of the turbidity peak during the disturbance are used to determine the degree of fouling.

CRITERIA FOR CLEANING

Introduction

Water companies need to have a general measure of fouling in connection with the cleaning of water main systems. The resuspension potential has been used for some years in The Netherlands. The size and duration of the turbidity peak that relates to the resuspension potential is the measure used to indicate the degree of fouling of a water main system. However, because

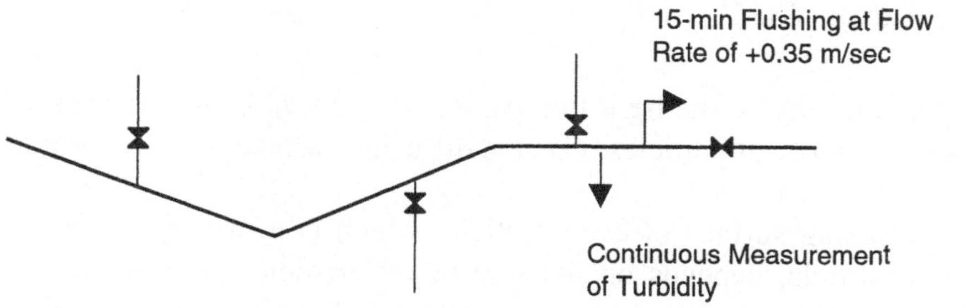

Figure 4.6 The principle of the Kiwa method

various methods are employed, no general criteria can be stated for indicating that a water main system is fouled.

In addition, because hardly any information is available about the turbidity level at which customers begin to complain, the turbidity level at which cleaning is required has not yet been defined.

Effect of Turbidity Measurement on the Resuspension Potential Measurement

Working in cooperation with Kiwa Water Research, Dutch water companies have developed simplified methods for measuring the resuspension potential to determine the degree of network fouling and the effect of cleaning operations. The simplified method consists of continuously measuring turbidity for a 1-hour period. Because the simplified method generates separate measurements in different periods, it is important for the turbidity meter to make reproducible measurements.

In addition, water companies in The Netherlands use different types of turbidity meters for the measuring, and it is known that different turbidity meters can produce different measurement values. Consequently, the results obtained with different turbidity meters cannot be simply compared.

In the past, Kiwa has generated a large volume of data from monitoring systems that use SIGRIST turbidity meters. Water companies want to know the degree to which measurements made using other types of turbidity meters correspond with those made using the Kiwa monitoring system.

Principles of Turbidity Measurement

Measurements obtained from different turbidity meters cannot be simply compared because turbidity is highly dependent on the nature of particles, the wavelength of the light used, and the turbidity meter's construction.

When light is shone into a solution that contains particles, some of the light is diffused, absorbed, or both, and the rest passes through. If the intensity of the light that passes through is compared with that measured when no particles are present (e.g., in clean water), it is possible to determine the amount of light that is diffused (Gregory and Nelson 1984).

The amount of light that passes through depends on the nature and size of the suspended particles, the concentration, and the wavelength of the light path according to the following formula:

$$I = I_0 \exp(-NCL)$$

where I is the intensity of the light that passes through, I_0 is the intensity of the light beam that enters, N is the number of particles, C is the diffusing surface of each particle, and L is the optical wavelength.

The diffusing surface of each particle, which is a measure of the total amount of light diffused by a particle, depends on the size of the particle, the refractive index, the shape and surface structure of the particle, and the wavelength of the light that enters.

The color of the particles affects the level of the light's diffusion. Because colored particles absorb light, diffusion by these particles is lower.

Moreover, dissolved substances may color drinking water. Absorption of diffused light can reduce the signal, resulting in a lower turbidity measurement. By measuring diffused light through an angle of 90° and by measuring the light that passes through, it is possible to compensate for the effects of the color and fouling of the measure cell.

Turbidity meters are calibrated using standard formazine solutions with a known turbidity. When calibrated, different turbidity meters may produce the same turbidity measurement. However, if solutions with different particles are then measured, the turbidity reading may differ among instruments.

Comparison of Turbidity Meters in the Laboratory

In The Netherlands, two types of turbidity meters are commonly used in the methods for measuring resuspension potential: the SIGRIST KT65 and the Dr. Lange Ultraturb turbidity meters. The degree to which they measure turbidity differs. Kiwa uses the SIGRIST KT65 turbidity meters in its monitoring systems. Because of their measuring principle, dimensions, weight, and price, these meters are not suitable for the simplified method.

The Dr. Lange Ultraturb and the SIGRIST KT65 turbidity meters have been compared with each other in a series of measurements (Slaats 2000). Table 4.1 briefly describes the two meters.

In a laboratory situation, the turbidity meters were compared on the basis of continuous turbidity measurements in water mains. The turbidity meters were set up in series and SIGRIST turbidity meters were placed before and after the Dr. Lange turbidity meter. Two other types were involved in the test (X,Y); their results will not be given in this report. Figure 4.7 shows the arrangement of the turbidity meters. The water flow through the turbidity meters was approximately 80 L/hr and was set using a flowmeter. Measurement data were read once per minute and stored in a data logger.

Measurement Results From the Laboratory

The turbidity of the drinking water in the laboratory generally shows minor variations. The sediment in the pipe was resuspended by turning on a fire hose during the test period. The velocity and changes in velocity in the supply pipe were not known.

The various instruments were also used to measure the turbidity of a formazine solution with a known turbidity. To enable this, the solution was placed in the different instruments.

Figure 4.8 shows the signal from the turbidity meters for the increase in turbidity when the fire hose was turned on. Turning on the fire hose resulted in resuspended sediment in the supply pipe and increased water turbidity. The height and shape of the peak was found to depend on the

Table 4.1
Description of the two compared turbidity meters

	Dr. Lange Ultraturb	SIGRIST KT65
Measuring principle	ISO* 7027	White light
Light source	Infrared light-emitting diode (IR LED)	Tungsten lamp
Wavelength light source	880 ± 30 nm	360–2,200 nm
Volume of measure cell	±0.1 L	0.25 mL
Measurement range	0.001 to 1,000 ftu	0–2, 0–5, 0–10, and 0–20 ftu
Resolution	0.01 ftu	1% of the set scale
Calibration	Factory method Formazine basis 0 and 8 ftu	Factory method Formazine basis 0 and 8 ftu
Cleaning	Present	Compensation of signal
Velocity	Minimum 12 L/hr	
Maximum pressure (25°C)	5 bar	
Design	Dry and wet section in one; electronic section is splash-proof	Dry and wet section separate

*International Organization for Standardization.

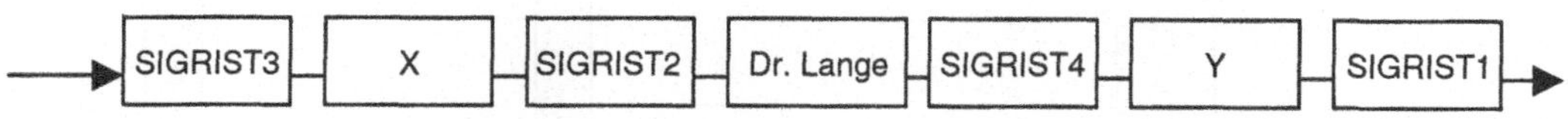

Figure 4.7 Arrangement of turbidity meters during the test period

type of turbidity meter. The Dr. Lange turbidity meter measured a turbidity of more than 2 formazine turbidity units (ftu) for a long period. The SIGRIST turbidity meters also measured an increase in turbidity, but the figures measured never rose above approximately 0.6 ftu.

The turbidity of a standard solution was measured to determine which turbidity meter measured the correct level. The standard solution consisted of a formazine solution with a turbidity of 1.5 ftu. The laboratory measurements were made using a Dr. Lange LPT4 Ultraturb turbidity meter. This was calibrated in accordance with the instructions of Sterlab (the Dutch association for laboratory accreditation) and is based on ISO 7027. The solutions with high turbidity were measured using only the SIGRIST3 and SIGRIST2 turbidity meters, and these gave the lowest and highest measurement values for continuous measurements during the entire period. Figure 4.9 shows the results. The Dr. Lange turbidity meter gave a signal that was approximately 30% higher than the value of the laboratory meter. The SIGRIST turbidity meters generally gave a figure that was comparable to that of the laboratory meter; the individual values were lower and higher.

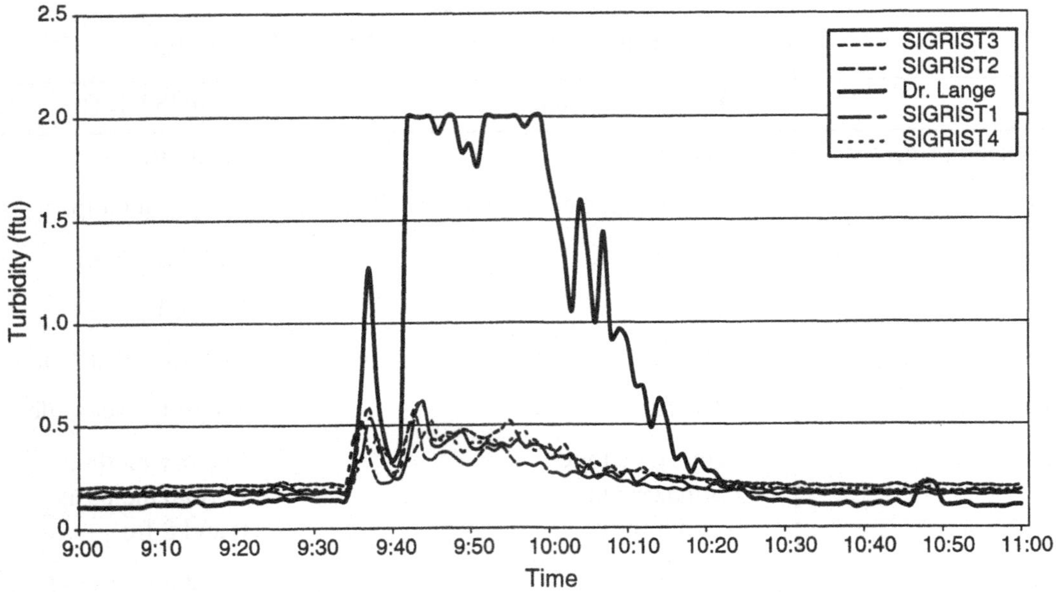

Figure 4.8 Turbidity during the introduction of a disturbance measured with several turbidity meters

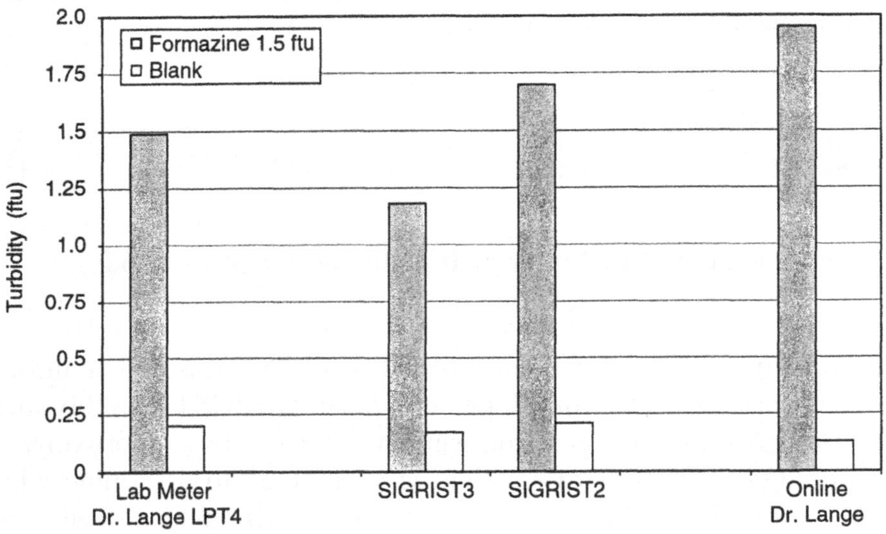

Figure 4.9 Measurement values with a formazine solution with the laboratory turbidity meter and the online turbidity meters

The turbidity meters clearly give different turbidity measurements. The differences are partly explained by the light source that was used. The Dr. Lange turbidity meter uses an IR LED. This diode transmits IR light of a fairly narrow bandwidth.

The Dr. Lange laboratory meter's light source is a halogen lamp. The light from a halogen lamp passes through a filter that permits light of 800–1,000 nm to pass. The light falls on a diode that is sensitive to light between 0–920 nm. This means that light with a wavelength of between 800 and 920 nm is detected, which corresponds with approximately 860 nm ± 60 nm. In comparison with the Ultraturb, the laboratory meter has more diffused light because of the optics and the cuvette that are used.

Conclusions From the Laboratory Comparison

The Dr. Lange Ultraturb and SIGRIST turbidity meters measure comparable turbidity patterns in the turbulent solution. However, because the absolute figures measured are at different levels, measurement results from the different turbidity meters cannot be compared with each other.

When specifying the criteria for cleaning, it is necessary to state the measuring instrument that was used when defining the criteria.

Comparison of Results of Simplified Methods in the Field

Various companies have given their own shape to the resuspension potential measurement by using various sampling points and various turbidity meters. The choice of sampling point can affect the outcome of the measurements. Moreover, measurements made using different turbidity meters cannot be simply compared with each other.

To make a comparison of the different measuring methods used, three different methods were used to take measurements at a large number of companies (Kreukniet 2001). This resulted in comparable measurement values for fouling in the water main system.

Implementation in the Field

To compare the outcomes of the various methods, the OPM, PWN, and Kiwa methods were deployed simultaneously at each location. The individual methods are described in the Simplified Resuspension Potential Methods section. Figure 4.10 is a diagram of the implementation methods for the measurements of this kind. The figure shows a pipe section that was isolated by closing valves so that a definite flow direction was established. A tap-in point was installed on the pipe in the vicinity of a fire hydrant. During the measurement, turbidity meters were placed on the fire hydrant and the tap-in point.

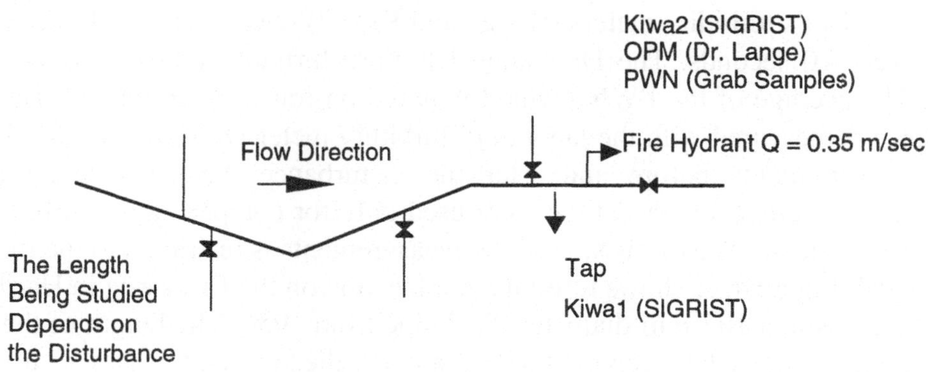

Figure 4.10 Diagram of arrangements at measurement locations

The OPM measurement was made using a Dr. Lange Ultraturb turbidity meter connected to a fire hydrant. Two Kiwa monitoring systems, using SIGRIST turbidity meters, were also installed, one on the fire hydrant in series with the Dr. Lange turbidity meter (Kiwa1) and the other at the tap-in point (Kiwa2). During the measurements, grab samples were taken through the fire hydrant (i.e., the PWN method). The disturbance was prepared in accordance with a protocol, which is detailed in appendix A.

The zero situation for turbidity in the drinking water was determined about 10 minutes after starting a measurement. A disturbance of 0.35 m/sec was then introduced through a fire hydrant. The duration of the disturbance depended on the length of the isolated section and was, in this case, 15 minutes. After the disturbance, turbidity was measured for another 15 minutes.

Measurement Locations

The measurements were made in 42 locations at the following companies: Water Company Overijssel (WMO), Hydron Central Netherlands (HMN), Water Company Gelderland (WGld), Water Company Limburg (WML), Water Company East Brabant (WOB), and Water Company Drenthe (WMD). Table 4.2 shows details of the measurement locations. The velocity in the pipe was based on data from the water company and was not measured. The measurements were made in the period from July to August 2001.

Measurement Results From the Field

A comparison of the measurement values gives an insight into the differences between

- Measurements made through a tap-in point and through a fire hydrant (Kiwa1 and Kiwa2)
- Measurements made using two different turbidity meters (Kiwa2 and Dr. Lange [OPM method])
- A continuous measurement or measurements made by taking grab samples (OPM method and PWN method)

The measured turbidity values are displayed graphically in Figure 4.11 to allow the different methods to be compared. Furthermore, the turbidity average was determined during the disturbance (5 or 15 minutes) and for the duration of the PWN method (between 3 and 12 minutes). Kiwa's turbidity meters (Kiwa1 and Kiwa2) measure the turbidity continuously and record it every 10 seconds. The Dr. Lange Ultraturb turbidity meter records the results every 1 minute. The average of the PWN method is based on four measurements. The turbidity of the grab samples is measured with the laboratory turbidity meter Dr. Lange LPT4. The results of the turbidity measurements before and after the disturbance were not taken into account in determining the average values. These were used only for comparing the turbidity pattern of the various measurements. The discussion of the measurements is restricted to the three measurement examples used. Figure 4.11 shows turbidity at a location on the Brinkweg in Hardenberg (WMO). This location was in a 107-mm-diameter PVC pipe from 1956. The length of the isolated section was 500 m. The distance between the fire hydrant and the tap-in point was 17 m. The disturbance introduced at the fire hydrant was 11 m^3/hr. Table 4.3 shows the statistical averages of turbidity during the disturbance (between 3 and 12 minutes).

Table 4.2
Details of measurement locations

		WMO	HMN	WGld	WML	WOB	WMD	Total
Estimated velocity	<0.1 m/sec	1	6	Unknown	6	2	4	19
	0.1–0.5 m/sec	3	1	Unknown	1		3	8
	>0.5 m/sec	3		Unknown		5		8
Pipe material	A–C	7	4	4	4	4	1	20
	PVC		2	3		1	3	9
	Cast iron		1		3	1	3	8
Diameter	85 mm	1						1
	100 mm		5	3	7	2	4	21
	110 mm	6	2	3		1	2	14
	125 mm					1		1
	150 mm					2		2
	160 mm						1	1
	200 mm			1		1		2

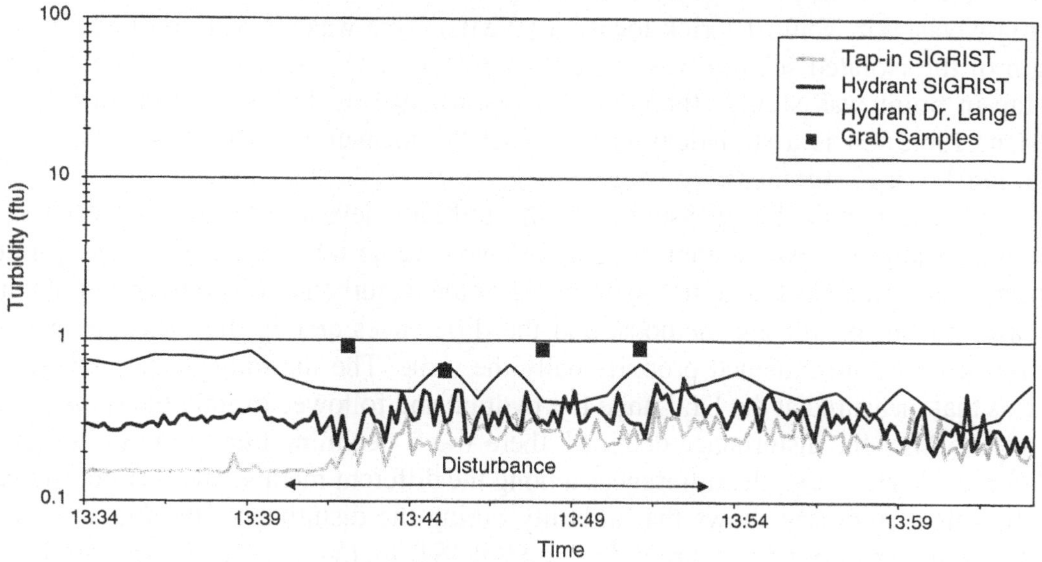

Figure 4.11 Turbidity measurements in a clean pipe

Table 4.3
Turbidity during an introduced disturbance in a clean pipe

	Average (ftu)	Minimum (ftu)	Maximum (ftu)
SIGRIST tap-in point	0.3	0.2	0.4
SIGRIST meter at fire hydrant	0.4	0.2	0.5
Dr. Lange Ultraturb meter at fire hydrant	0.5	0.4	0.7
Grab samples measured in laboratory (Dr. Lange LPT4)	0.8	0.7	0.9

Figure 4.11 shows that the turbidity levels differed from each other after the measuring equipment was started. The lowest level was measured at the tap-in point, and the Dr. Lange fire hydrant measured the highest turbidity level. After introducing the disturbance, it took approximately 2 minutes for a slight increase in turbidity to be observed. Turbidity at the SIGRIST tap-in point and the SIGRIST fire hydrant increased slightly and presented an agitated turbidity picture. Turbidity measured at the Dr. Lange fire hydrant did not change. Turbidity of the grab samples was higher than that seen with the other measuring methods. Only slight differences in the measured turbidity were apparent during the disturbance. There is hardly any effect on turbidity of the increased velocity for the grab sample method or the Dr. Lange fire hydrant, which indicates the pipe is clean.

The statistical summary shows that turbidity during the disturbance (the measure of fouling) was 0.25 ftu for the tap-in point and 0.35 ftu for the SIGRIST fire hydrant. The figures for the Dr. Lange fire hydrant and the grab samples were 0.5 ftu and 0.8 ftu, respectively.

Measurement in "Slightly Fouled Pipe." Figure 4.12 shows the level of turbidity at the Hertog Alvastraat, Venlo-Blerick location (WML). This was a 100-mm-diameter A–C pipe. The length of the isolated section was more than 300 m. The distance between the fire hydrant and the tap-in point was shorter than 5 m. A disturbance of 11 m^3/hr was introduced at the fire hydrant. Table 4.4 is a statistical summary of the measurement results during the disturbance (between 3 and 12 minutes).

Likewise with this measurement, the turbidity levels differed from each other after the measuring equipment was started. Again, the lowest value was measured at the tap-in point and the highest value at the Dr. Lange fire hydrant. After the disturbance was introduced, the turbidity level increased for all measuring methods, and the differences among the measurements made on the different equipment remained proportionally the same. The turbidity of the grab samples was the same as that measured at the Dr. Lange fire hydrant and followed exactly the same pattern seen with that meter. After the disturbance declined, there was a declining trend in turbidity levels for all the methods used. Here, too, the differences among the different methods were proportionally the same. The statistical summary shows that turbidity during the disturbance (the measure of fouling) was 1.9 ftu for the tap-in point and 3.8 ftu for the SIGRIST fire hydrant. The figures for the Dr. Lange fire hydrant and the grab samples were 10.7 ftu and 10.2 ftu, respectively.

Measurement in "Fouled Pipe." Figure 4.13 shows the turbidity measurement for the location Portengen 47 (HMN). This is a 110-mm-diameter PVC pipe. The length of the isolated section was 480 m, and the distance between the fire hydrant and the tap-in point was 200 m.

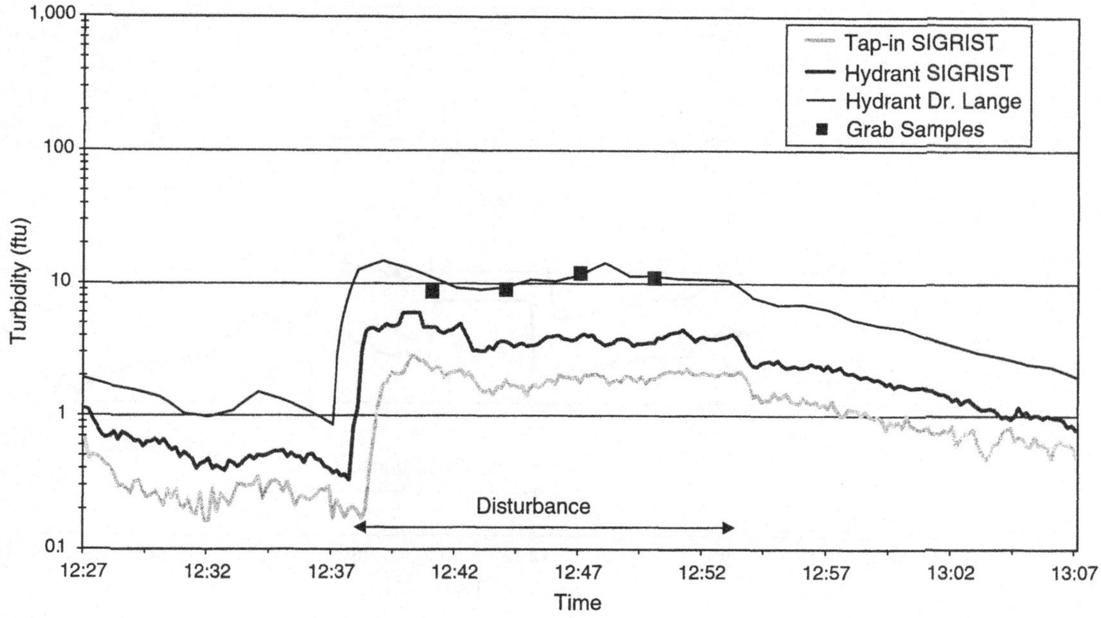

Figure 4.12 Turbidity measurements in a slightly fouled pipe

Table 4.4
Turbidity during an introduced disturbance in a slightly fouled pipe

	Average (ftu)	Minimum (ftu)	Maximum (ftu)
SIGRIST tap-in point	1.9	1.4	2.3
SIGRIST fire hydrant	3.8	3.1	5.0
Dr. Lange fire hydrant	10.7	9.0	14.3
Grab samples (Dr. Lange LPT4)	10.2	8.7	12.0

A disturbance of 11 m^3/hr was introduced at the fire hydrant. Table 4.5 is a statistical summary of turbidity during the disturbance (between 3 and 12 minutes).

Immediately after starting the equipment, turbidity was comparable with the two examples discussed—lowest at the tap-in point and highest at the Dr. Lange fire hydrant. After the disturbance was introduced, the turbidity level increased for all measuring methods. The turbidity level at the SIGRIST fire hydrant reached a higher value than the maximum of the measurement range (>6 ftu). Halfway through the disturbance, turbidity fell abruptly at the fire hydrant, but turbidity at the tap-in point remained at the same high level. The large distance between the two measuring points (MPs) probably caused this. It is possible that the fouling situations in the section between the tap-in point and the fire hydrant were different from those in the section before the tap-in point. The turbidity levels of the grab samples were more or less the same as that of the Dr. Lange fire hydrant and followed the pattern of the continuous meter precisely. After the

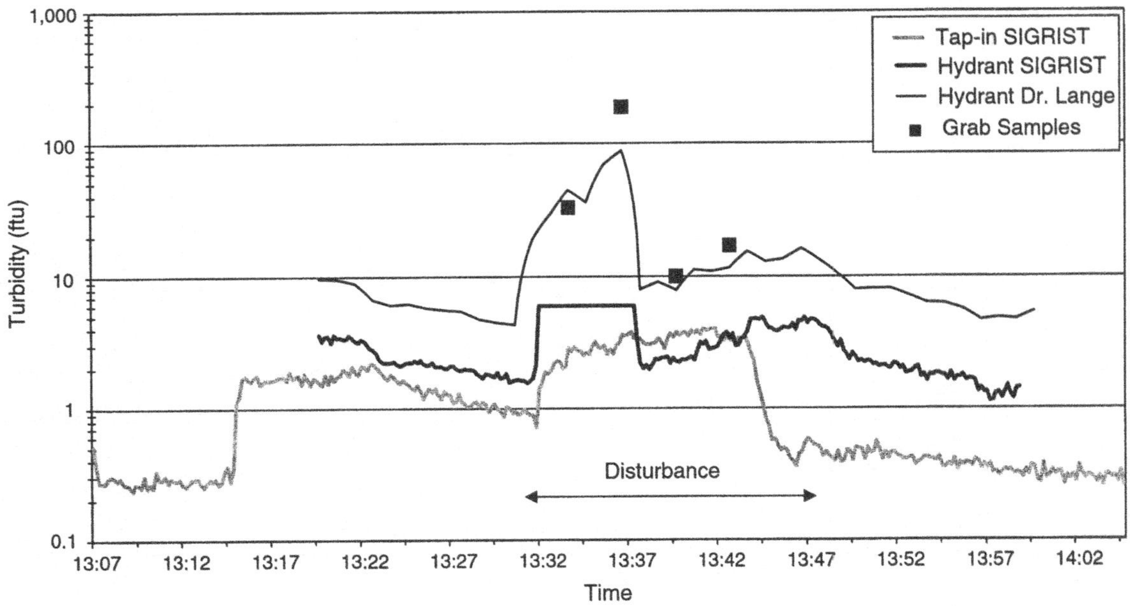

Figure 4.13 Comparative measurements at a fouled location

Table 4.5
Turbidity during an introduced disturbance in a fouled pipe

	Average (ftu)	Minimum (ftu)	Maximum (ftu)
SIGRIST tap-in point	3.3	2.5	4.0
SIGRIST fire hydrant	4.0	2.0	6.0
Dr. Lange fire hydrant	29.7	7.9	88
Grab samples (Dr. Lange LPT4)	62	10	188

disturbance declined, there was a declining trend in turbidity levels for all the methods used; however, the decline at the tap-in point was more abrupt.

The statistical summary shows that turbidity during the disturbance (the measure of fouling) was 3.3 ftu for the tap-in point and 4.0 ftu for the SIGRIST fire hydrant. The figures for the Dr. Lange fire hydrant and the grab samples were 29.7 ftu and 62 ftu, respectively. Note that the turbidity meter at the SIGRIST fire hydrant could measure only up to 6 ftu, so this value was used to determine the average. The actual average during the disturbance would, therefore, have been higher.

To allow comparisons among all the measurements, the level of fouling has been expressed as a figure for each method and location. The investigators decided to do this in accordance with the method used by PWN, because for all the investigated methods, turbidity values were obtained in a time interval between 3 and 12 minutes. In the PWN method, a

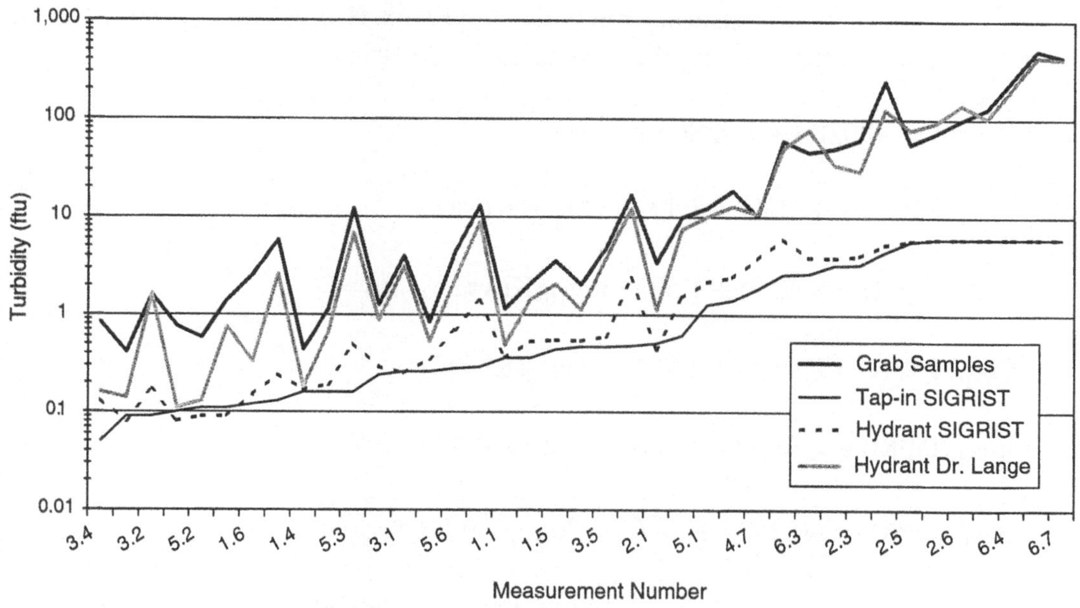

Figure 4.14 Graph of the average values per method for all locations

location's level of fouling is determined on the basis of the arithmetical average of the four samples and is expressed in formazine turbidity units.

Figure 4.14 shows the average values per method for all locations. Kiwa1 results are arranged in ascending order. The figure shows that the four methods produced systematically different results—the PWN method generally produced the highest measurement values, the OPM method the second highest, the Kiwa2 method the third highest, and the Kiwa1 method the lowest.

In the measurement range shown in Figure 4.14, approximately 0.1 ftu, the Kiwa2 method (fire hydrant), usually produced higher measurement values than the Kiwa1 method (tap-in point). From this we can deduce that, in that measurement range, measuring at a fire hydrant gives higher values than measuring at a tap-in point. This may be caused by material being released from the fire hydrant, by a different flow pattern in and around the fire hydrant, or both. The differences between the measurement points were not caused by a difference in the equipment because the two SIGRIST turbidity meters were identical and were calibrated together.

When comparing the measurement values at the fire hydrant, the Dr. Lange turbidity meter systematically registered higher measurements than the SIGRIST turbidity meter. This picture corresponds with that obtained from laboratory tests made in the past (Slaats 2000). The difference may have been partly caused by the fact that the Dr. Lange turbidity meter uses IR light for its measurements, whereas the SIGRIST turbidity meter uses visible light.

The OPM and PWN methods produce an almost identical turbidity pattern during measurements. However, differences occur in the *size* of the turbidity values. The number of measurement values used to calculate the average may cause this difference.

Statistical Relationship Between the Results of the Measuring Methods

Conversion formulas were worked out to allow the results of one measuring method to be translated into those of another. The formula was then worked out using regression analysis for each combination of two measuring methods. Appendix B presents the basis for this statistical analysis. The relationships among the investigated measuring methods is shown as

$$\text{Kiwa2} = -0.0386 + 1.4286 \cdot \text{Kiwa1}$$

$$\log_{10}(\text{PWN}) = 1.1081 + 1.1316 \cdot \log_{10}(\text{Kiwa1})$$

$$\log_{10}(\text{OPM}) = 0.9256 + 1.583 \cdot \log_{10}(\text{Kiwa1})$$

$$\log_{10}(\text{PWN}) = 0.9138 + 1.1861 \cdot \log_{10}(\text{Kiwa2})$$

$$\log_{10}(\text{OPM}) = 0.6429 + 1.4600 \cdot \log_{10}(\text{Kiwa2})$$

$$\log_{10}(\text{OPM}) = -0.3524 + 1.1651 \cdot \log_{10}(\text{PWN})$$

Table 4.6 summarizes the turbidity values obtained using this formula to convert a turbidity value measured with Kiwa1 (SIGRIST tap-in point) into values obtained from other measuring methods.

Translation of Measurement Results Into Criteria for Cleaning

Up to now, water companies have had no uniform criteria to indicate whether a pipe is clean or fouled. For example, Kiwa uses a limiting value of 2 ftu (measured using the Kiwa method), above which a pipe is considered fouled. However, PWN uses a limiting value of 60 ftu (measured using the PWN method). Table 4.7 shows the criteria used thus far for the Kiwa and PWN methods.

Table 4.6
Comparison of turbidity values obtained under different measuring conditions

Kiwa1 (SIGRIST tap-in point, ftu)	Kiwa2 (SIGRIST fire hydrant, ftu)	OPM (Dr. Lange fire hydrant, ftu)	PWN (grab samples, ftu)
0.5	0.7	2.8	5.9
1	1.5	8.4	13
2	2.8	25	28
2.5	3.5	36	36
5	7.1*	108	79

* In practice, the SIGRIST turbidity meter can only measure up to 6 ftu.

Table 4.7
Criteria for determining the degree of fouling of the water main system

Pipe fouling	PWN method—average turbidity (ftu) of 3-, 6-, 9-, 12-min samples	Kiwa method—resuspension potential (ftu)
Clean	≤20	≤1
Slightly fouled	>20 and ≤60	>1 and ≤2
Fouled	>60	>2

Table 4.8
Criteria for degree of fouling converted from the Kiwa method

Pipe fouling	Kiwa1 (SIGRIST tap-in point, ftu)	Kiwa2 (SIGRIST fire hydrant, ftu)	OPM (Dr. Lange fire hydrant, ftu)	PWN (grab samples, ftu)
Clean	≤1	≤1.5	≤8	≤13
Slightly fouled	>1 and ≤2	>1.5 and ≤3	>8 and ≤25	>13 and ≤28
Fouled	>2	>3	>25	>28

Using the Kiwa method, a location is considered fouled if the turbidity figure is greater than 2 ftu for more than 2 hours. The average turbidity value during that period, then, is at least 2 ftu. The average turbidity value during the disturbance (the first 15 minutes) will also be at least 2 ftu.

Using the criteria for the Kiwa method and the conversion models referred to previously, it is possible to indicate the limiting values for the investigated measuring methods. Table 4.8 summarizes the limiting values.

The values for fouling converted on the basis of the Kiwa method are lower than the standard figure PWN adopts for cleaning. Although the stated values remain rather arbitrary, they do give a good indication of a location's fouling.

In the OPM method, a Dr. Lange turbidity meter was used. Of course, other turbidity meters can be used in the OPM method, as long they fulfill the specifications given in Table 4.1. However, the specific sets of criteria for cleaning must be derived for a specific turbidity meter and, of course, for local circumstances.

Panel Session to Estimate Level of Complaints

A panel was brought in to help make an estimate of the turbidity level at which customers start to complain. The results of the panel, along with accumulated experience, were used as the basis for establishing criteria for avoiding complaints from customers by determining when the water main system should be cleaned.

During a workshop with 90 participants from Dutch water companies and Kiwa employees, Kiwa conducted an exploratory investigation to determine the turbidity level at which

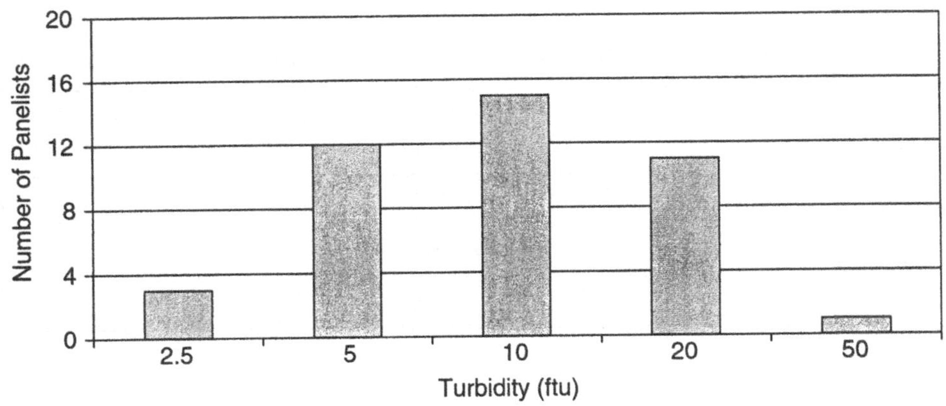

Figure 4.15 Turbidity levels at which panel members would complain to the water company

the panel would complain to the water company. Forty of the 90 participants voted on the turbidity level at which they would complain to the water company.

For the experiment, turbid solutions were made in a range of 2.5, 5, 10, 20, and 50 ftu, and turbidity was measured using a Dr. Lange laboratory meter. The measurement figure obtained was the same as that obtained using a Dr. Lange turbidity meter of the kind used in the OPM method.

The turbid parent solution consisted of drain water from a cleaning operation carried out in cast-iron pipes at Nieuwegein (HMN). Drinking water, also from Nieuwegein, was used to prepare the diluted solution.

The solutions were presented to the panel members in six white, identical porcelain dishes containing a few liters. During the session, the solution was stirred slowly using a magnetic stirrer to prevent particles from sinking. The participants, who were unaware of the absolute turbidity of the solution, were asked to indicate at which stage they would complain.

Figure 4.15 gives an overview of the turbidity level at which complaints would be made. The average turbidity level at which members of the panel would complain was 12 ftu. This value corresponds reasonably well with the value for the resuspension potential for slight fouling, as shown in Table 4.8.

MEASURING PROGRAM USING THE RESUSPENSION POTENTIAL METHOD

Fouling and Cleaning Cycle

Over time, water mains will contain an increasing degree of fouling. When this reaches a certain critical level, the system must be cleaned. Turbidity monitoring, or the filter color method, can be used as an indicator in the resuspension potential method for determining when the system needs to be cleaned. The number of customer complaints can also be used to determine when the system needs cleaning. Refouling will occur again after cleaning, and Figure 4.16 is a schematic illustration of the fouling and cleaning cycle.

Regular measurement and keeping of records of complaints will give insights into the rate of refouling. Supposing a linear relationship between pipeline fouling and time, the time of the next cleaning operation can be determined by extrapolating the results of M1 and M2.

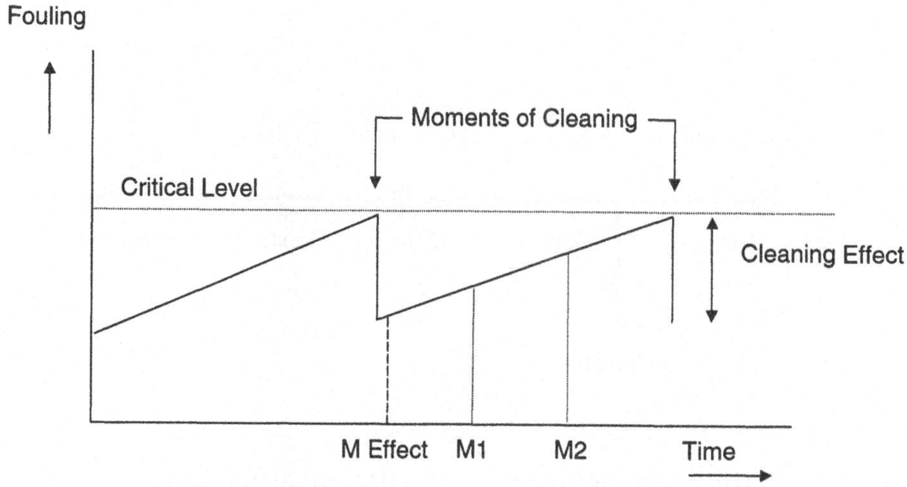

Figure 4.16 Schematic representation of the effect of fouling and cleaning in a pipeline over time, with respect to a critical level (M stands for measurement of the fouling of the network)

Measurement after a cleaning action (M effect) gives an insight into the effectiveness of the action.

Applying the Resuspension Potential Method

The resuspension potential method is used for different purposes. For example, it is commonly used in connection with determining

- The fouling mechanism of a distribution area (section measurement)
- The degree of fouling of a distribution area (the resuspension potential measurement or RPM)
- Specific measurements

Section Measurement

To verify a change in water quality in trunk mains, a section measurement is obtained by placing a series of monitors (which measure turbidity constantly) along the main, as shown in Figure 4.17. Changes in the water quality along the section give an insight into the degree of fouling. By preference, the first monitor is located at the outgoing main of the pumping station to account for the effect of changes in water quality caused by the pumping station. One monitor is placed at the beginning of the route and another at the end. Depending on the length of the route and the available equipment, one or more monitors are placed along the main being monitored. They are located especially at spots where a decreasing water quality is to be expected, such as downstream of highly corroded pipes.

Decreasing turbidity indicates sedimentation along the route (at least during the monitoring period). Continuous sedimentation will, in the long term, result in pipeline fouling. Especially during hydraulic disturbances, the sediment may be eroded and consequently foul the downstream area. If the turbidity remains constant throughout the entire section, we assume that

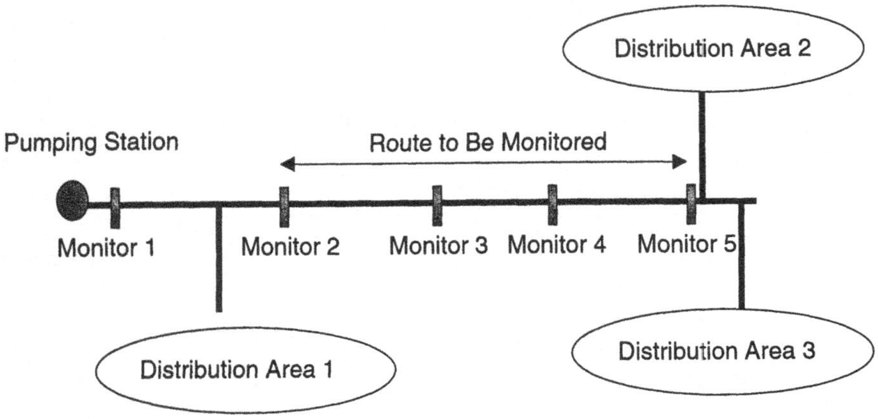

Figure 4.17 Schematic representation of a section measurement

no sedimentation or uptake will take place along the route. Increasing turbidity during transport indicates that sediment is being picked up during transport along the route. Increasing turbidity can be detected in the following cases:

- If the water quality of the pumping station has recently been improved. In that case the pipes will be cleaned naturally, which does not generally lead to water quality problems.
- If a part of the pipeline consists of corroded cast iron, and iron pick-up is occurring
- If the pipeline's capacity to accommodate sediment is exhausted because of earlier sedimentation. This increases the risk of water quality problems for the customer.

Fouling of the Distribution Network

By measuring the water quality at a limited number of locations in a distribution area, we can predict the degree of fouling of that particular area. Methods such as the OPM are suitable for performing measurements of this kind, which can give an indication of fouling after measuring for a short period. These measurements can be performed consecutively and in a short time period (about 1 to 2 hours per measurement).

Monitoring water quality in different distribution areas delivers information on the degree of fouling per distribution area. When used in combination with, for example, the number of customer complaints, priorities for establishing a cleaning program can be set. To obtain a representative view of the state of fouling, various locations must be measured. Measurements can be based on obtaining a general indication of the state of fouling. In that case, various more or less representative pipelines must be selected, based on criteria such as velocity or pipe material. The measurement can also be based on a worst-case situation. In that case, monitoring is performed at the areas where the worst conditions are expected (corroded pipes, low velocities, high retention times). For an indication of the number of measurements, see the section on the cost of measuring versus flushing in chapter 8.

By repeating the monitoring, preferably at the same locations, information is gained about the rate of fouling of the distribution area. As shown in Figure 4.16, consecutive monitoring gives

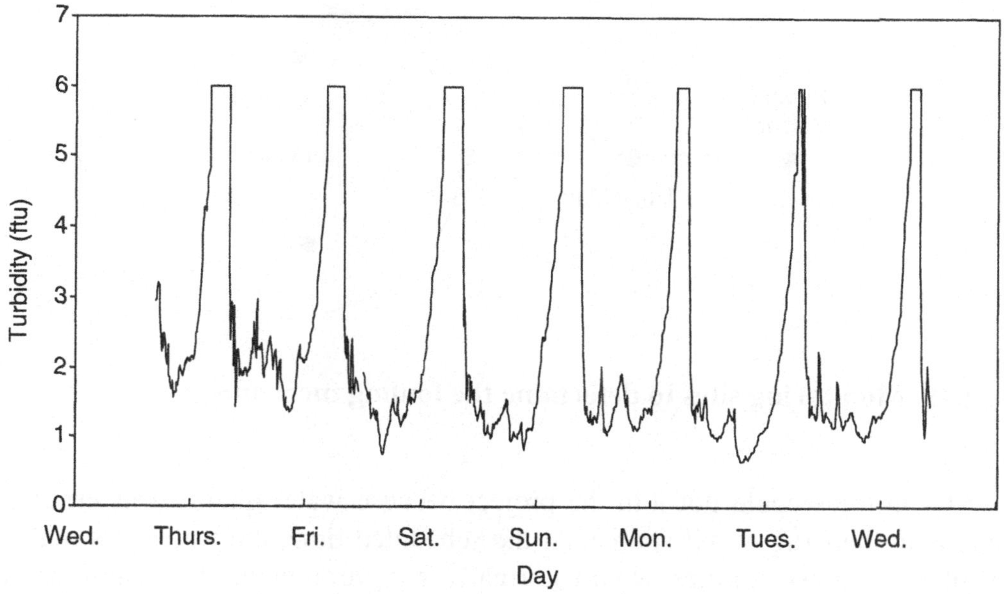

Figure 4.18 Typical example of a turbidity pattern at a corroded cast-iron pipe

an indication of the required cleaning interval. Moreover, monitoring before and after a cleaning operation allows its effect to be quantified.

Specific Measurements

To determine the location and impact of a specific local source of fouling, measurements can be performed specifically in the neighborhood where the fouling is thought to be located. The need for measurements of this kind may be dictated by a large number of turbidity incidents in a concentrated area, or if a hot spot is found, by indicative measurements. Turbidity is continuously measured over a few days at a number of locations.

A typical pattern that can be observed during this type of measurement is that of iron release from a corroding cast-iron pipe, as shown in Figure 4.18. The pattern of increasing turbidity during the overnight hours between 11:00 p.m. and 7:00 a.m. results from a process in which corrosion products are constantly being absorbed by the water. During low velocities at night, the water remains stagnant and turbidity increases. When the water is first used in the morning, turbidity decreases abruptly because the pipeline's contents have been replenished.

RESULTS OF RESUSPENSION POTENTIAL METHOD IN THE UNITED STATES

In the second stage of the project, several techniques were used to test the hypothesis and to evaluate methods. One pilot location was selected for discolored water risk assessment in the United States, at Cincinnati Water Works in Ohio. The trial in the United States was meant to demonstrate the value of the monitoring techniques for discolored water risk assessment.

Other utilities in the United States were unable to cooperate in the project at the scheduled time. At some utilities, special projects such as this were not being conducted because of the terrorist attacks on Sept. 11, 2001, and the resulting heightened security of their water systems.

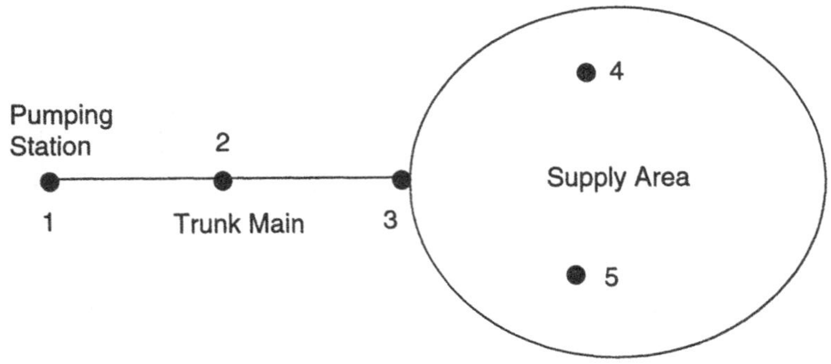

Figure 4.19 Monitoring sites to determine the fouling mechanism

Other water utilities could not join the project because water quality was erratic, as a result of fluctuations in their raw-water source. At the scheduled time, the water was not a representative sample of what these utilities would normally see throughout the year—largely because of drought conditions and the increased reliance on alternative water sources. Other utilities did not have the capability to conduct measurements.

The results of the measurements in Cincinnati are compared with the results in The Netherlands. If the trends and patterns are the same, the resuspension potential method can be used to predict discolored water events in the United States, just as it is used in The Netherlands. Therefore, two types of measurements were carried out at Cincinnati Water Works:

1. Measurements to determine causes of fouling of the network (fouling mechanism)
2. Measurements using the resuspension potential to determine the degree of fouling

Fouling Mechanism

To determine the fouling mechanism, five monitoring sites in Cincinnati were selected between the Western Hills pumping station and the supply area, as shown in Figure 4.19.

The locations were

1. Western Hills effluent pumping station
2. Main trunk, Grand Avenue between Sandwell Avenue and Utah Avenue (400-mm diameter)
3. Main trunk, Grand Avenue north of Lehman Road (400-mm diameter)
4. Fairbanks South of Price Avenue (400-mm diameter)
5. Chateau North of West 8th Street (150-mm diameter)

All locations were monitored for 1 day, except location 3, which was monitored for 2 days. The monitoring sites were equipped with three identical turbidity meters with data loggers (Dr. Lange), powered by 24-V batteries. The turbidity meters were connected to a main by a hose and a ferrule. The flush volume was approximately 200 L/hr.

Location 1: Western Hills Pumping Station

Location 1 is situated 9 m from the last pump in the outlet of Western Hills pumping station. This location was monitored on Oct. 23 and 24, 2001. The results showed a relatively high average turbidity level of the water leaving Western Hills pumping station (above 2 ftu), with several turbidity changes occurring during the day. Based on the experience of Cincinnati Water Works, the results obtained seem unreliable, so they are not given here. Measuring of the turbidity of grab samples in August 2002 resulted in turbidity of about 0.5 ftu.

Location 2: Grand Avenue Between Sandwell Avenue and Utah Avenue

Location 2 is situated on the 400-mm-diameter main trunk at Grand Avenue. It was monitored on Oct. 23 and 24, 2001, and the results are shown in Figure 4.20.

Average turbidity during the monitoring period was 0.18 ftu. However, the turbidity pattern was unstable, with many short-term turbidity spikes, with a maximum of 0.9 ftu. These spikes indicate the presence of sediment, which resuspends during short-term increases in velocity.

A typical turbidity pattern for active cast-iron corrosion was not detected during the monitoring period, indicating that corrosion layers were stable. The results show that this location was contaminated with sediment.

Location 3: Grand Avenue North of Lehman Road

Location 3 is situated on the end of the main trunk at Grand Avenue and was monitored from Oct. 23–25, 2001. Data from this location are not presented here because the turbidity measurement went offline at this location, probably because of air bubbles in the measurement cell.

Location 4: Fairbanks South of Price Avenue

Location 4 is situated on a main at Fairbanks south of Price Avenue and was monitored on Oct. 24 and 25, 2001. Data from this location are not presented here because the turbidity measurement went offline at this location, probably because of air bubbles in the measurement cell.

Location 5: Chateau North of West 8th Street

This location, situated on a 300-mm-diameter main at Chateau North of West 8th Street, was monitored on Oct. 24 and 25, 2001. The results are shown in Figure 4.21. Average turbidity at this location during the monitoring period was 4.0 ftu. Turbidity fell to 2 ftu after 7:30 p.m. Turbidity at this location was high, especially for a trunk of this size.

A typical turbidity pattern for active cast-iron corrosion was not detected during the monitoring period, which indicated that the corrosion layer was stable. Sediment was also picked up between locations 2 and 5.

From the results, we can conclude that the sediment load presented to the supply area accounted for the fouling mechanism in the distribution area of Western Hills pumping station. This can be seen from the relatively high fluctuations in the turbidity pattern at the inlet of Western Hills pumping station and the increase of turbidity between locations 2 and 5. Neither of the cast-iron mains showed signs of active corrosion.

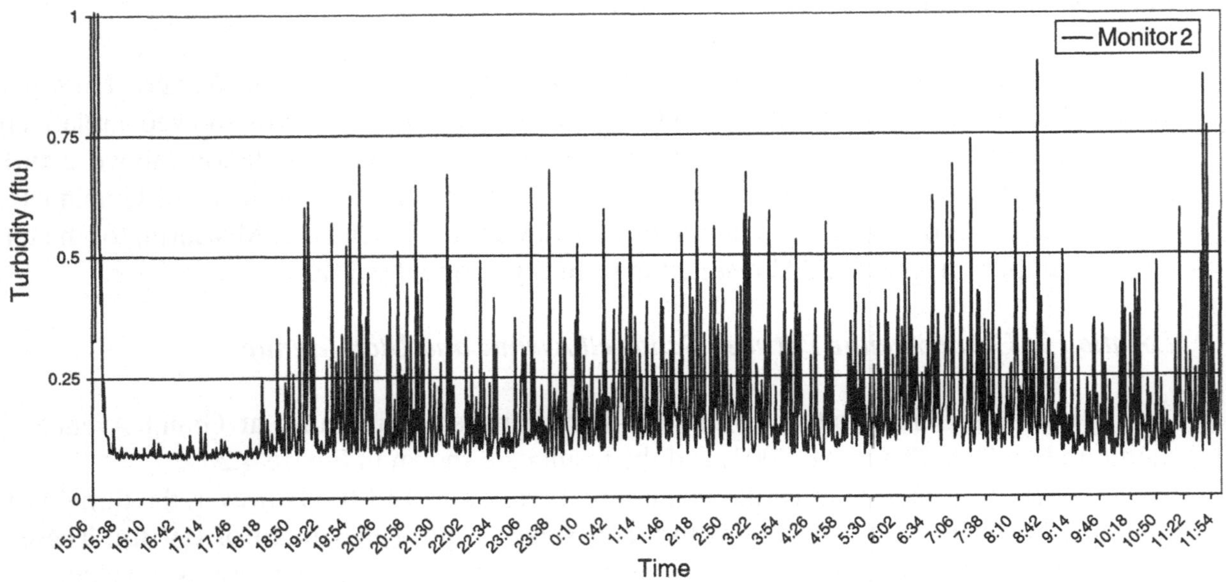

Figure 4.20 Turbidity at location 2, the main trunk at Grand Avenue, Oct. 23–24, 2001

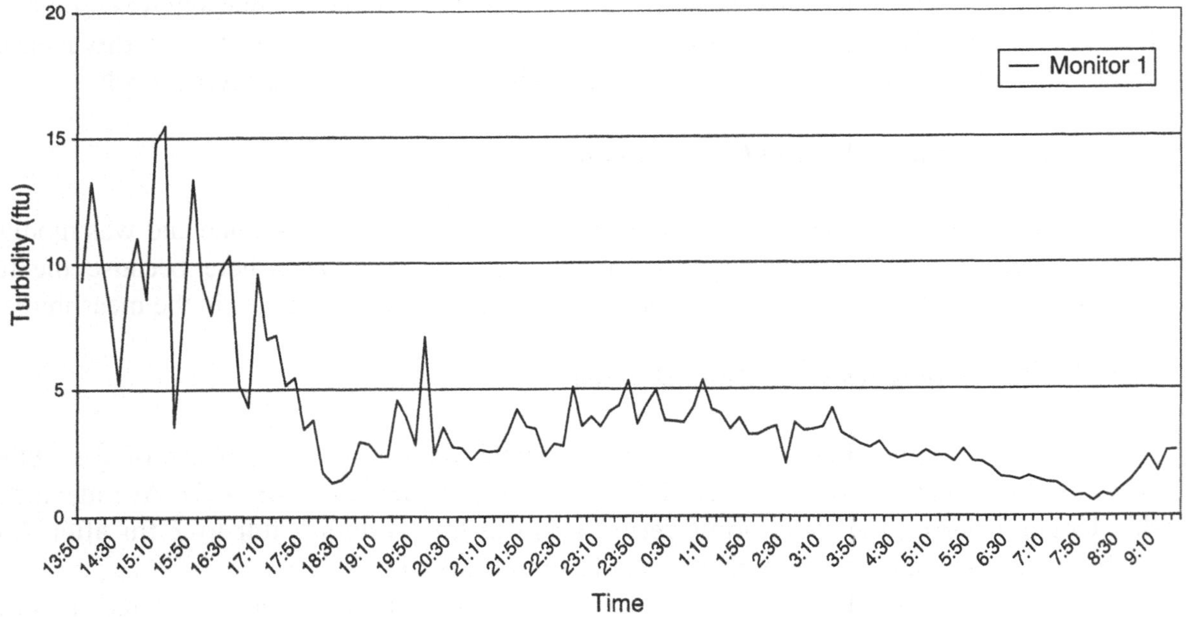

Figure 4.21 Turbidity at location 5, Chateau North of West 8th Street, Oct. 24–25, 2001

In addition to the conclusion on the fouling mechanism, some experience was also gained about the experimental aspect of the resuspension method. While monitoring turbidity at five locations, the measurement went offline at two locations (locations 3 and 4). Air bubbles in the measure cell probably caused this. High pressure in the main, which had to be reduced before the monitoring began, caused pressure at the monitor to be almost atmospheric. At that time, dissolved gases expanded and created little gas bubbles (this was observed at the location of Warsaw Avenue, near Wilder Avenue). These bubbles became entrapped in the measure cell and influenced turbidity, a phenomenon that occurred at every location. The higher flow through the measure cell prevented the monitor at the other three locations from going offline.

Air bubbles can be caused by

- Air leakage at the suction side of the pump, resulting from incorrect or worn out seals
- Supersaturation of oxygen, caused by entrapped air after repairs
- Shifts in the chemical equilibrium of hydrogen carbonate and carbon dioxide, resulting from a change in pH

It also is clear that special attention needs to be paid to the presence of air bubbles in the network, the analysis of turbidity data, and the fouling mechanism, because turbidity is not related only to sediment in the network.

Short Resuspension Potential Method

The RPMs were carried out in three different ways:

1. Online turbidity measurements at the flushing point
2. Online turbidity measurements at a service tap close to the flushing point
3. Grab samples from the flushing point at certain times

These measurements take about 1 hour to perform.

The measurements were carried out at six sites located in the same supply area where the fouling mechanism was monitored. Some of the sites are not listed here because the results from only two sites are presented—location 7, Considine Avenue north of Basset Road, and location 8, Purcell Avenue south of Lehman road. A flow disturbance was introduced at the hydrant, and a diffuser was used to measure the increased velocity. However, the diffuser was unable to measure low velocities. For this reason, the introduced disturbances were somewhat higher than the required 0.35 m/sec.

Location 7: Considine Avenue North of Basset Road

The main on Considine Avenue (150-mm diameter) was isolated, and one turbidity monitor was installed on the main at the end of Considine, near Basset Road. The other monitor was installed at the hydrant on Considine, south of Basset Road. The normal velocity in the main at this location is somewhere around 0.05 m/sec at this time of the day.

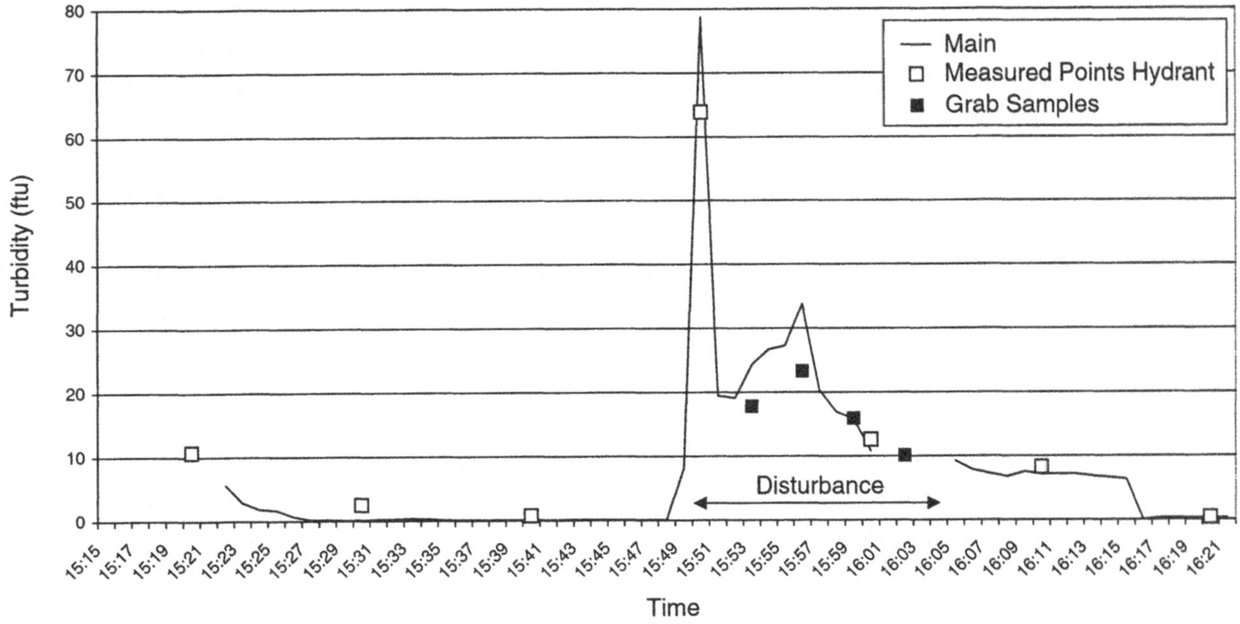

Figure 4.22 Turbidity resuspension potential at Considine Avenue, Oct. 25, 2001

At 3:50 p.m., a velocity of 0.93 m/sec was introduced to start the disturbance. An installation fault in the monitor at the hydrant meant that measurements were made only once every 10 minutes instead of every minute. Figure 4.22 shows the results. Right at the start of the disturbance, the turbidity at the main peaked to a maximum of 79 ftu. After a minute, turbidity dropped back to 20–30 ftu. After the disturbance was stopped, turbidity remained at 7 ftu. At 5:17 p.m., it dropped back to 0.3 ftu, as a result of the replenishment of the main's contents. The monitor and grab samples at the hydrant measured the same pattern. Maximum turbidity at the hydrant was 64 ftu. Judging by the high level of disturbance, this location was slightly contaminated by resuspendable sediment.

Location 8: Purcell Avenue, South of Lehman Road

Purcell Avenue (150-mm diameter) was isolated down to Warsaw Avenue. The first monitor was installed on the main at the beginning of Purcell Avenue, south of Lehman Road. The other monitor was installed at the hydrant where Purcell Avenue and Lehman Road intersect. Normal velocity in the main at this location at this time of the day was somewhere around 0.13 m/sec.

At 10:43 a.m., a velocity of 0.89 m/sec was introduced to start the disturbance. The results are shown in Figure 4.23. At the start of the disturbance, turbidity at the main increased slowly and reached its maximum of 43 ftu 10 minutes after the disturbance started. When the disturbance was stopped, the turbidity dropped to 7 ftu. The turbidity measurement at the hydrant followed the same pattern. Turbidity reached a spike of 40 ftu only at the start of the disturbance. The grab samples also showed the same pattern (with the highest turbidity, 29.8 ftu, at the end of the disturbance). The high turbidity indicated that this main was slightly contaminated with resuspendable sediment.

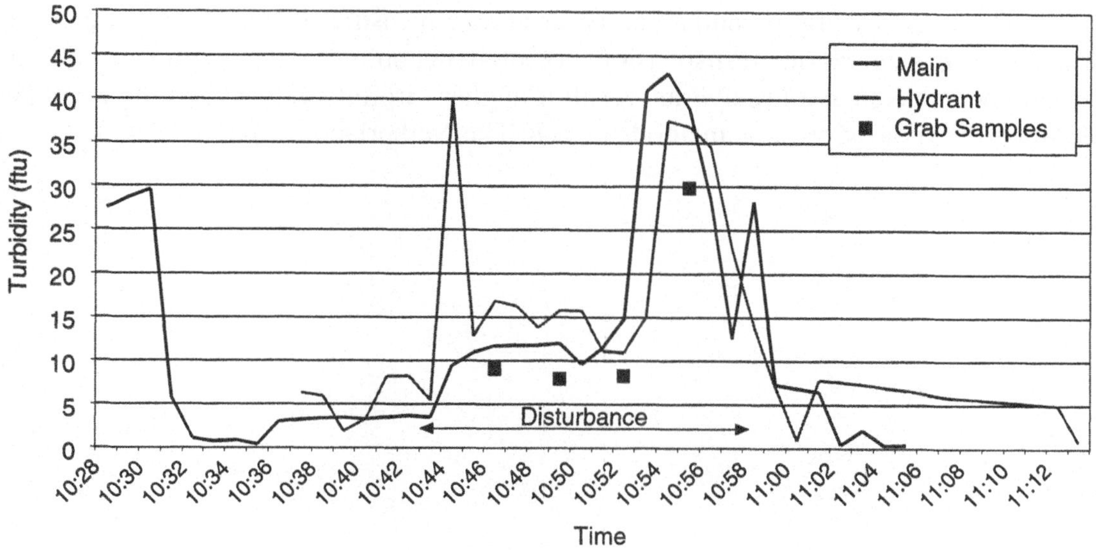

Figure 4.23 Turbidity resuspension potential at Purcell Avenue, south of Lehman Road, Oct. 26, 2001

At the end of these experiments, we could conclude that it seems possible to use the resuspension potential method under the conditions that occur in the main system at Cincinnati Water Works in Ohio. Some technical modifications and improvements appear necessary.

The results would indicate that the Western Hills pumping station presents a continuous sediment load to the supply area. The turbidity increases in the network, which indicates that sediment is being picked up during transport.

There was no indication of cast-iron corrosion.

In spite of the high velocity of the disturbance used during the resuspension potential method, the resuspension potential at the measured locations in the supply area was limited. The supply area was slightly contaminated with resuspended sediment.

In addition to the conclusion on the fouling mechanism, some experience was gained about the experimental aspect of the resuspension method. Some measurements failed because of air bubbles in the measure cell, which influenced turbidity. It is clear from this that special attention must be paid to the presence of air bubbles in the network, the analysis of turbidity data, and the fouling mechanism, because turbidity is not related only to sediment in the network in such cases.

CONCLUSIONS

Discolored water complaints from customers can serve as a measure of fouling in a water main system. Complaint records can be very useful in determining the degree of fouling of a distribution area.

Fouling of water mains can be measured using various methods that are based on the resuspension potential principle. All the methods are suitable for measuring the degree of fouling in a pipe.

The resuspension potential can be used as a measure of fouling in a water main system. However, the size of the resuspension potential depends on the method used; for example, measurements taken at a fire hydrant result in higher turbidity values than measurements taken at a tap-in point. For the various methods used in The Netherlands, refer to Table 4.8.

CHAPTER 5
DISCOLORED WATER RISK MANAGEMENT

As discussed in the last chapter, various techniques are available for determining the presence of discolored water. The techniques yield information from various perspectives on fouling in the water main system and on the likelihood of discolored water complaints. Complaints indicate that discolored water is actually being seen at the tap, meaning that the water company is too late in taking measures because customers are already affected.

Given that sediment is the main cause of discolored water, it can be prevented or eliminated in three ways:

1. Prevent particles from entering the network.
2. Prevent suspended solids from settling.
3. Remove sediment from mains effectively and in due time.

STEP 1: PREVENT PARTICLES FROM ENTERING THE NETWORK

When particles are prevented from entering the network, no resuspension will take place and discolored water will not occur; this, then, is the optimal way to prevent discolored water.

As discussed in chapter 2, water quality is an important factor in the formation of particles. To recap, the processes responsible for generating discolored water may be

- Processes leading to particles entering the distribution system:
  - Directly from the treatment plant
  - From corrosion of water mains or domestic installations
  - From inadequate installation or maintenance work
- Processes leading to particle formation within the distribution system by physical, chemical, and biological processes (ongoing coagulation, formation, and abrasion of biofilm)

It is necessary to examine whether modifying or optimizing the treatment process will result in less sediment getting into the water main system. If economically and socially feasible, the treatment plant should be modified.

Modifying the water quality at the treatment plant can also affect particle formation within the distribution system by physical, chemical, and biological processes. Therefore, the corrosiveness and the biostability of the water must be examined.

STEP 2: PREVENT SUSPENDED SOLIDS FROM SETTLING

When the velocity of the water reaches a certain value every day, suspended solids will not (or will only briefly) be subject to sedimentation. The solids will be removed in such a way that customers will not see any signs of turbidity. The water will be supplied as produced. One way of combating discolored water is to design self-cleaning water main systems. Because of its branched construction and relatively small pipe diameters, a self-cleaning distribution network has a unidirectional flow and higher velocity than a conventional network.

STEP 3: REMOVE SEDIMENT FROM MAINS EFFECTIVELY AND IN DUE TIME

Removing sediment from the water main system almost always helps to prevent discolored water complaints. The possibility of cleaning having an adverse effect occurs only in the case of cast-iron pipes. The cleaning cycle can be determined using the techniques described in chapter 4.

The resuspension potential method allows for objective determination of the need to clean a network. Once the necessity of cleaning has been demonstrated, an effective way of cleaning should be chosen.

The next three chapters offer more details on the three steps outlined in this chapter.

CHAPTER 6
STEP 1: PREVENT PARTICLES FROM ENTERING THE NETWORK

INTRODUCTION

Two main sources of sediment in the network are generally recognized:

1. The substances present in the water leaving the treatment plant
2. Corrosion of cast-iron mains in the distribution network

The water quality at the treatment plant may be deficient in terms of iron and manganese removal, corrosivity for cast iron, and biostability. Most processes that influence water quality also influence each other.

The quality of and procedures used for operations and maintenance are of paramount importance to the control of water quality in both production and distribution. As discussed previously, water quality is an important factor in the formation of particles entering the network. Water treatment can be optimized to restrict the number of particles entering the network, and water composition can be optimized or treated to prevent cast-iron pipes from corroding.

MONITORING PARTICLES ENTERING THE NETWORK

The Filtrate Scan

Limiting the volume of solid substances leaving the treatment plant can reduce the amount of sediment in the water main system. To bring about this volume reduction, the amount of particulate matter should be monitored. The filtrate scan has been developed in The Netherlands as a tool for measuring particles in the drinking water leaving the treatment plant (Siegers and IJpelaar 2002).

The filtrate scan, which consists of a set of techniques for measuring and characterizing particles in the treatment plant, can be used to

1. Determine whether discolored water complaints can be traced to particles in the water leaving the treatment plant
2. Ascertain the cause of the "high" concentration of fouling substances

Figure 6.1 is an overview of the measuring techniques used and the purpose of the measurement. The filtrate scan is used to analyze both the final filtration stage in the treatment plant and the outgoing clean water.

The filtrate scan uses various methods to measure the presence of particles:

- Turbidity. The discharge of particles from the treatment plant is related to the operations and the filter's serviceable life. To obtain an impression of particle discharge, turbidity is continuously measured for a certain period. In turbidity measurements, particles of >0.2 μm are measured.

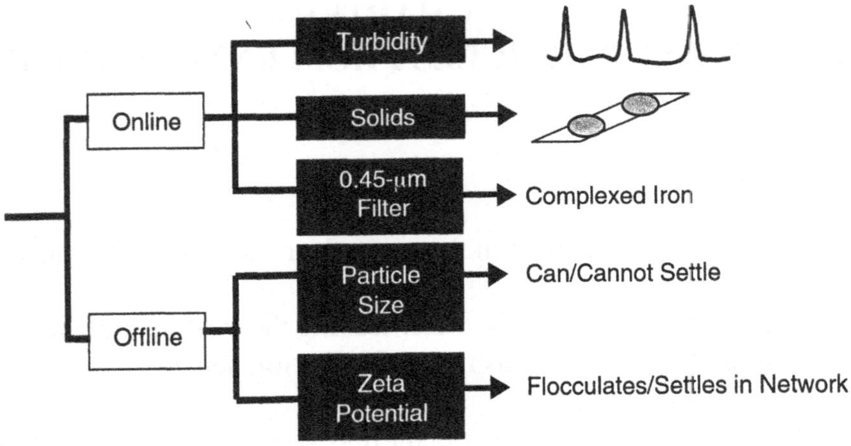

Figure 6.1 Techniques available in the filtrate scan

- Particle size (coarse) and nature of particle. An indication of the percentage of particles larger than 0.45 μm in the suspended substance can be obtained by filtering the water using 0.45-μm membrane filters. The measurement is made automatically at a constant pressure. The material collected on the filters is analyzed for iron (Fe), manganese (Mn), calcium (Ca), magnesium (Mg), and nonpurgeable organic carbon.
- Particle count and particle size. Dynamic light diffusion is used to analyze the particle size of the suspended substance in the filtrate. The technique reveals the presence of particles in the water with a size of >2 μm. The particle counter distinguishes among four fractions: 2–3 μm, 3–5 μm, 5–8 μm, and 8–100 μm.

  With the assumed particle-size distribution, the number of particles is inversely proportional to particle size. Therefore, there are more 2-μm particles than 3-μm, more 3-μm than 5-μm, and more 5-μm than >8-μm.
- Particle charge (zeta potential). The charge on the surface largely determines the coagulative properties of colloidal particles. The zeta potential is used to determine the charge on the surface of the suspended substance in the filtrate. This information makes it possible to say whether particles that leave the treatment plant are able to flocculate or settle.

  Generally speaking, a particle with a charge of < -15 mV or $> +15$ mV is (in theory) stable and does not coagulate. Particles with a charge between –15 and +15 mV tend to coagulate. If the value is close to zero, the likelihood of flocculation increases.

The filtrate scan can determine dissolved iron and manganese that leave the plant and have the potential to be oxidized and settle in the distribution system. These compounds can be analyzed in the water after filtration over a 0.05-μm filter. All compounds passing the filter are supposed to be in solution. The iron is probably complex iron (attached to organic compounds like humic acids) and the manganese is not properly oxidized to manganese dioxide (MnO_2) in the rapid sand filter. Both the iron complex and the Mn^{2+} can be oxidized in the network and precipitate.

Table 6.1
Overview of treatment steps at Noordbergum water treatment plant

	Treatment steps	Filtrate scan
Step 0	Obtaining of water	
Step 1	Aeration	
Step 2	Primary filtration	
Step 3	Softening	
Step 4	pH correction	
Step 5	Aeration	
Step 6	After filtration	Turbidity monitor "after filtration"
Step 7	Clear water vault	Turbidity monitor "clear water" Particle-size distribution

Case Study: Noordbergum Treatment Plant (Nuon)

Water companies can use the filtrate scan for various purposes. The Nuon water company, for example, used the scan to determine if there was a relationship between the operation of the Noordbergum treatment plant and the fouling of the water main system in the supply area (Siegers and IJpelaar 2002). The results of the study gave insights into the relationship between the cost of improving the treatment plant and any possible reduction in the frequency of water main system cleaning.

Measurements Using Filtrate Scan

During the examination of the Noordbergum treatment plant operation, turbidity measurements and particle counts were taken. The combination of turbidity measurements and particle counts showed a relationship between the number of particles in the water after the filters. The treatment of the water at Noordbergum consists of several steps, outlined in Table 6.1. Turbidity is measured after steps 6 and 7. Figure 6.2 shows the turbidity after the filter, as well as the turbidity of the clear water. Comparable patterns can be observed.

Figure 6.3 shows the turbidity of the clear water in relation to the particle count (total number of particles). We can also see in the figure that there is a strong correspondence between turbidity and the number of particles. Figure 6.4 illustrates the relative particle-size distribution during the measuring period at Noordbergum.

Additional information on the particles came from an analysis of the flushing water containing the sediment. Analyses of the composition of the water leaving the pumping station showed a correspondence with the composition of the flushing water including the sediment—a lot of calcium and little, if any, iron and manganese.

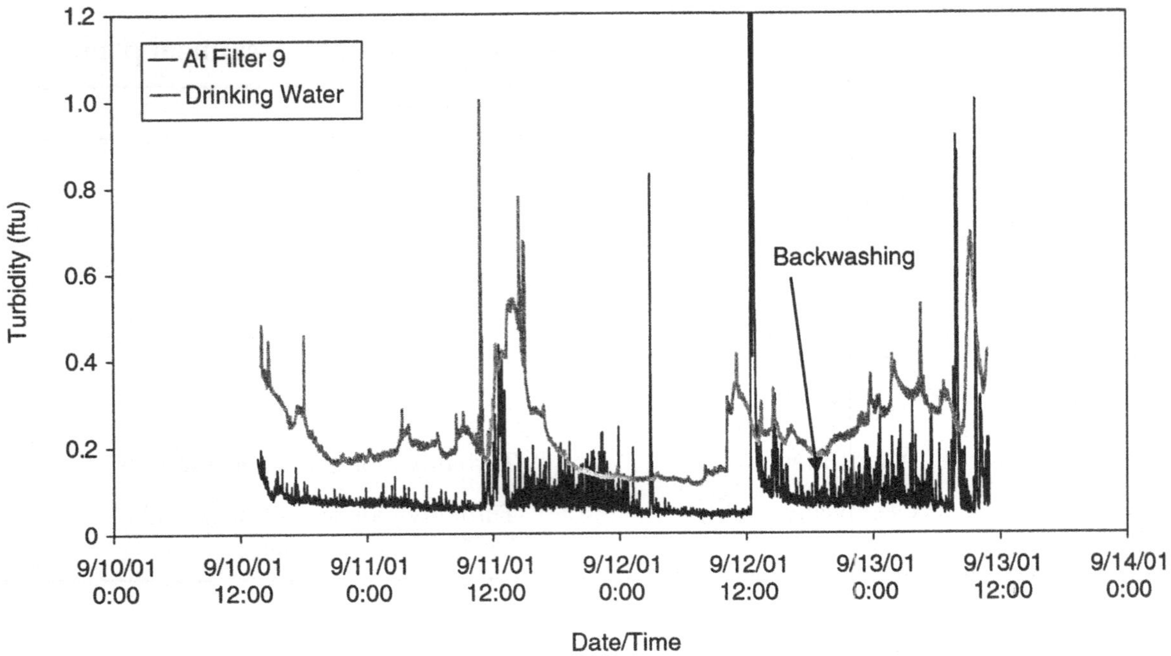

Figure 6.2 Turbidity of the water leaving at filter 9 (step 6) and the clear water (drinking water, step 7)

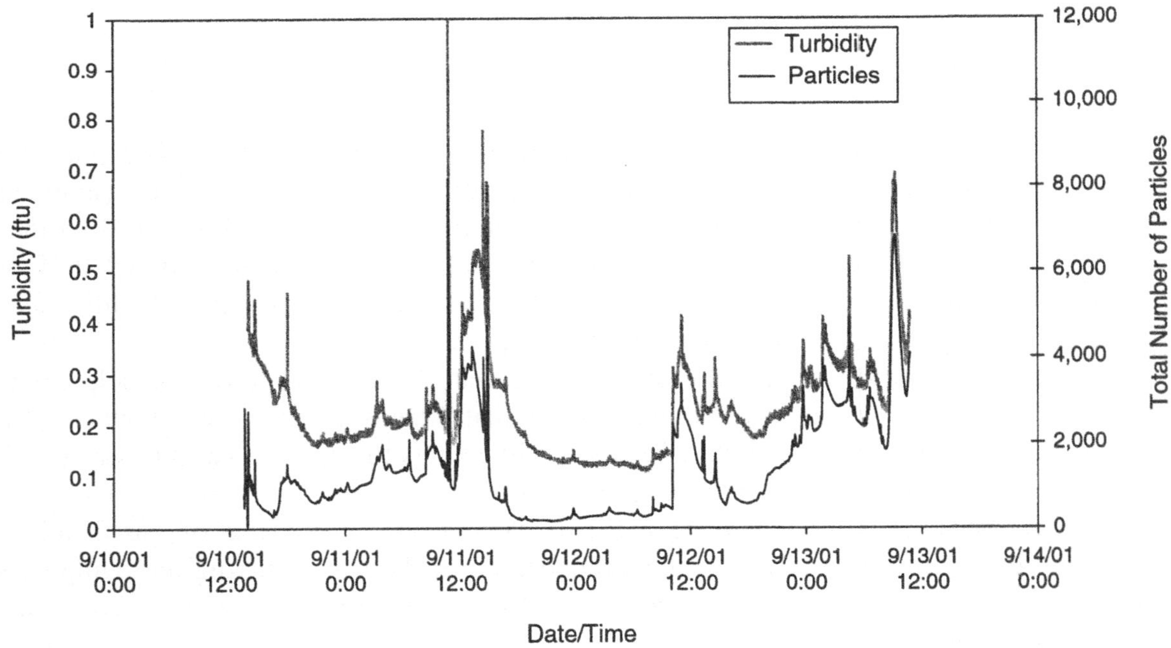

Figure 6.3 Turbidity versus particle counts in the clear water at Noordbergum

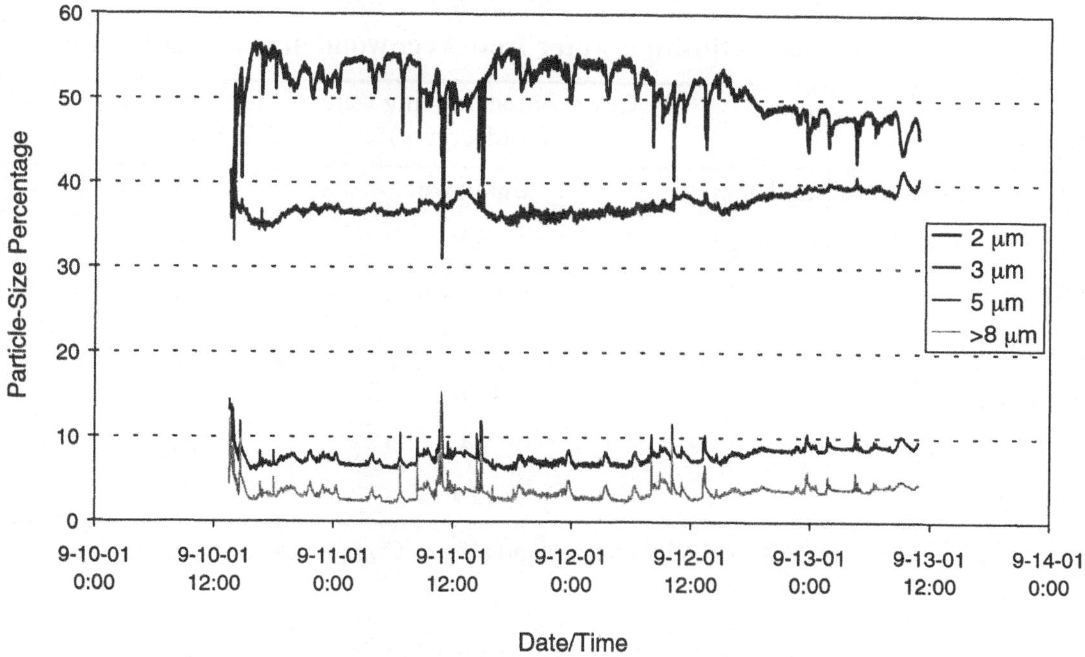

Figure 6.4 Relative particle-size distribution during the course of the measurements at Noordbergum treatment plant

Table 6.2 shows the analysis data from flushing activities in Veenwouden and Wouters-woude. Various pipes were involved in Veenwouden: 300-m cast iron (200 mm diameter), 200-m PVC (diameter unknown), and 150-m A–C (100 mm diameter). This leads to the conclusion that softening functions inadequately and that the secondary filters are incapable of catching the small calcium crystals that have not been removed.

Measurements using the filtrate of filter 9 show that the water contained particles with a zeta potential of approximately –10 mV, which means that these particles almost certainly coagulate. On the basis of the zeta potential, the particles in the filtrate have the potential to coagulate. This may mean that collisions between suspended particles may result in larger particles being formed, and those particles can settle and remain in the water main system.

Comparison of Flushing Costs

Using the filtrate scan has confirmed the inadequacies, of which the Nuon water company was already aware, in the Noordbergum treatment plant. The composition of the suspended substances in the flushing water is an especially clear indication that the sediment in the water main system consists mainly of calcium. The concentrations of iron and manganese in the sediment were extremely low. This implies that improving softening could lead to a significant reduction in sediment in the water main system. Besides calcium, the flushing water also contained varying concentrations of adenosine triphosphate (ATP), which is a measure of living biomass, suggesting that biofilms are present in the water main system. Biofilms may also contribute to sediment in the water main system.

Table 6.2
Analysis data on flushing water from Veenwouden and Wouterswoude

Parameter	Composition drinking water (Noordbergum)*	Flushing water (Veenwouden)	Flushing water (Wouterswoude)
Suspended substances (mg/L)	Unknown	270	117
Total organic carbon (TOC; mg C/L)	3.1	6.1	5.5
ATP (ng/L)	Unknown	690	190
Iron (mg/L)	<0.3	0	0
Manganese (mg/L)	<0.002	0.3	0.2
Calcium (mg/L)	44	134	97
Magnesium (mg/L)	12.1	11	11

**Water Quality Statistics* 2000, VEWIN (Netherlands Water Companies Association).

Case Study: Tull en 't Waal Treatment Plant (HMN)

The filtrate scan was also used to survey the situation at Tull en 't Waal treatment plant (Siegers and IJpelaar 2002). The measurements offer insight into the benefits that would result from limiting cleaning operations, if the treatment plant were to be improved.

The turbidity of the water after the filters, along with the clean water, was measured at Tull en 't Waal treatment plant for a specified period. Turbidity peaks were seen to occur regularly in the clean water. These could be traced to turbidity from the four filters. Figure 6.5 shows the turbidity from the filters at certain times in relation to the turbidity in the clean water.

Figure 6.6 shows the results of the turbidity and the total number of particles in the filtrate of filter 1. The results of the particle count corresponded with those of the turbidity measurements. Figure 6.7 shows the relationship to particle size over the duration of the measurements. The relationship shifted only during flushing, from small particles (2 μm) to large particles (>8 μm, indicated by an arrow).

The combination of turbidity and particle count showed that the particle size shifted from 2 μm to >8 μm during filter flushing. This may have resulted in larger particles being flushed out into the clean water, which could lead to sedimentation in the water main system.

An analysis of the flushing water, including the sediment, after flushing activities in the supply area of Tull en 't Waal treatment plant revealed that the water leaving the pumping station contained high concentrations of calcium and manganese and, to a lesser degree, iron. Table 6.3 shows the results of the analysis of the sediment from the water main system in the Tull en 't Waal supply area. We can see that the sediment consists mainly of manganese, iron hydroxides, and iron oxides, with a total of 120 mg/L of suspended substances. Because the manganese concentration in the raw water is a factor of 4 lower than the iron concentration, the high manganese concentration in the sediment-rich flushing water is striking. The water company was aware that the velocity achieved when backwashing the filters was insufficient for removing the manganese properly. Filtration during particle sedimentation in the pipes was incapable of removing all colloids.

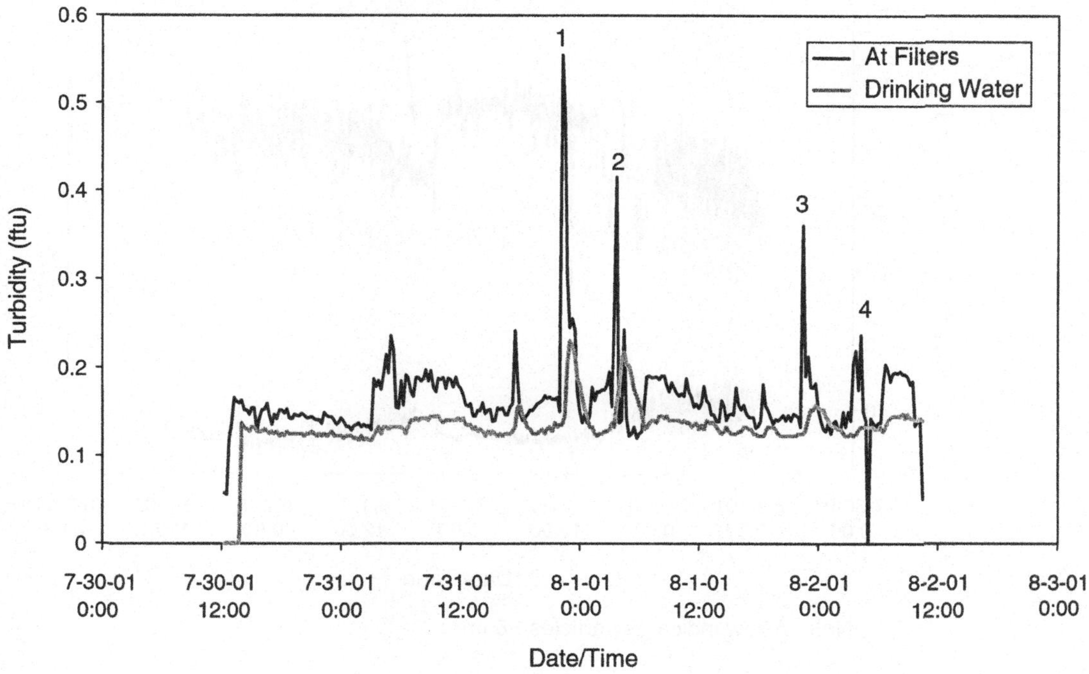

Figure 6.5 Relative contribution of the four filters of Tull en 't Waal to turbidity of the clean water

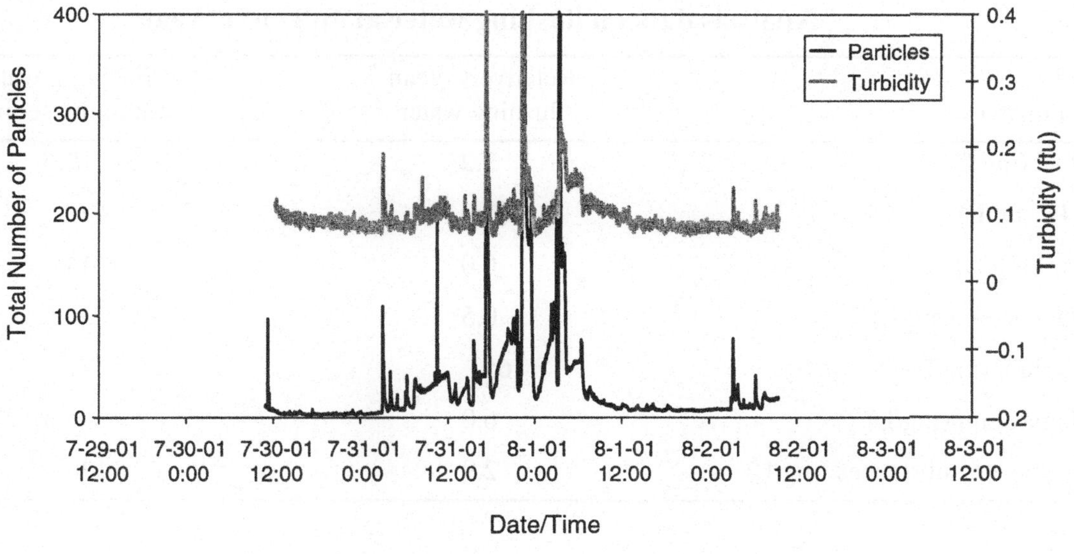

Figure 6.6 Turbidity versus particle counts in the filtrate of filter 1 at Tull en 't Waal treatment plant

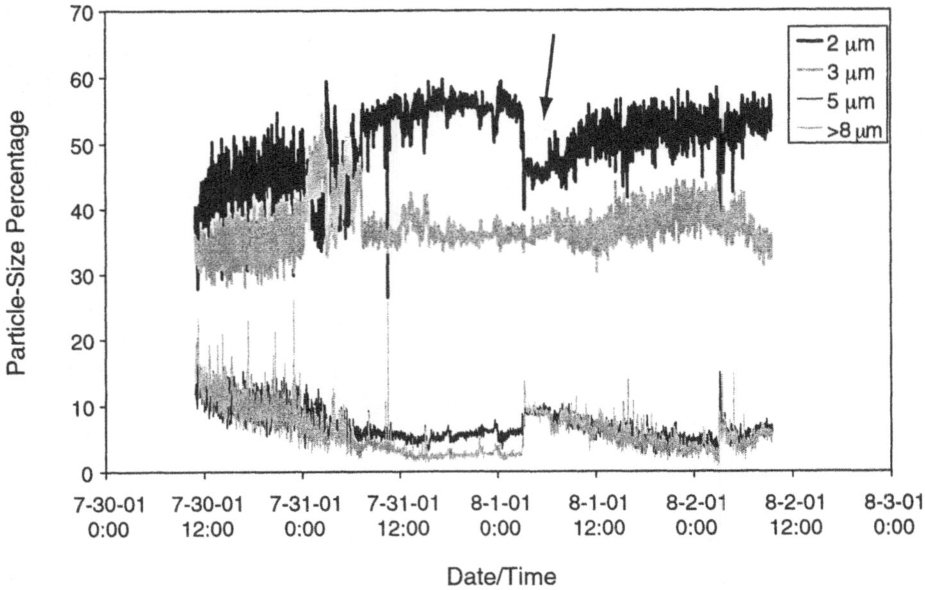

Note: Arrow indicates particles >8 μm.

Figure 6.7 Particle-size distribution during the course of the measurements at Tull en 't Waal treatment plant

Table 6.3
Analysis data on flushing water at Tull en 't Waal

Parameter	Relatively clean flushing water	Flushing water containing sediment
TOC (mg C/L)	2.2	9.5
ATP (ng/L)	31	470
Iron (mg/L)	0.0	14
Manganese (mg/L)	0.5	30
Calcium (mg/L)	65	77
Magnesium (mg/L)	6.0	6.3
Suspended substances (mg/L)	2.9	120

The water containing sediment also had higher concentrations of TOC and ATP, which indicate the presence of a biofilm in the pipe. The figure for ATP in relation to the surface area of the pipe is relatively high. However, it is not possible to make specific statements about the numbers of bacteria on the basis of the data.

Because the concentration of solids was too low, zeta potential measurements in the aerated and nonaerated raw water, the filtrate, and the clean water failed to produce any results.

On the basis of the survey and the scheduled modifications at Tull en 't Waal treatment plant, Kiwa advised HMN not to take any further measures in the treatment plant. Instead, the company was advised to concentrate on collecting measurement data from the existing treatment plant. In particular, Kiwa recommended that the profiles should be recorded for calcium, iron, and manganese throughout the filter's serviceable life and that various filter fractions should be characterized for the ability of particles to settle.

OPTIMIZING WATER COMPOSITION

Corrosivity of Water With Respect to Cast Iron

As previously mentioned, corrosion of cast-iron mains is seen as one of the main sources of discolored water. Controlling water quality to prevent or restrict corrosion of cast iron is one of the topics of research.

In Germany, the United Kingdom, and the United States, controllable pilot-scale pipeline systems are frequently used in discolored water investigations. Rig tests are carried out to study the relationship between water composition and corrosion, and between water composition and iron release. Technology Zentrum Wasser (TZW; Karlsruhe, Germany) uses continuous-flow systems, where a section of pipe material is exposed to water. Measurements are taken to determine weight loss or the changing chemistry of the test water.

The method developed at the Engler Bunte Institute (EBI) in Germany is worth mentioning (Kuch, Sontheimer, and Wagner 1983; Wagner and Kuch 1984). In this method, contact between ductile-iron pipe and the type of water under investigation is achieved within an isolated system. Oxygen consumption and iron concentrations are measured to supply information on corrosion rate and on iron release into drinking water, respectively. EBI used this method to determine the effect of various water parameters.

Results from the EBI study have shown that both corrosion rate and iron release increase when

- pH decreases, especially at pH values below the range from 7 to 7.5
- The amount of organic matter decreases
- Chloride and sulfate content increases
- Buffering capacity decreases. In this case, the quality of the corrosion layer (measured as percentage of iron built into the corrosion layer) decreases as well.
- Phosphate concentration decreases
- Strong fluctuations in water composition occur

Based on this information, a recommended water composition has been formulated in The Netherlands to reduce cast-iron corrosion (van den Hoven and van Eekeren 1988). To prevent corrosion of cast iron, we recommend that the pH should exceed 7.5, the total inorganic carbon (TIC) should exceed 0.5 mmol/L, and the corrosion index should be defined by $([Cl^-] + 2\,[SO_4^{2-}])/TIC$ expressed in mmol/L below 1.

Iron corrosion in drinking water distribution systems has been studied extensively in Sweden (Sander 1996). Although there is agreement that pH, calcium, and hydrogen carbonate are determining factors, the exact mechanisms are not fully understood. Drinking water is also treated in Sweden to prevent iron corrosion (Hedberg 1994). In the past, soft, low-pH water corroded unprotected iron mains. The resulting discolored water was dealt with by raising alkalinity with lime and carbon dioxide.

The effect of the corrosiveness of water is clearly demonstrated in several distribution areas in France, where discolored water problems are solved by adjusting the treatment of the drinking water in combination with pipe replacement (Boireau et al. 2000; Guerber, Magning, and Larigauderie 2000).

Treatment of the soft corrosive drinking water in Buenos Aires, where 14,000 km of old cast-iron piping is present, also seems to contribute to increasing the quality of the transported water (Cordonnier et al. 2000).

Much of the water in Brazil is corrosive, which results in water losses, discolored supplies, graphitization, reduction of flow capacity, and interference with water meters. These problems prompted pilot studies in Sao Paulo, resulting in recommendations to keep the pH between 7.0 and 8.0. Corrosion coupons are periodically inspected for tubercular corrosion development, as well as to determine corrosion rates (Cury 1994).

Recently, tests were performed in the United States to determine the effects of different water quality parameters on iron release, with a view to developing red-water control strategies (Clement et al. 2002). These were conducted by examining the release of iron at two U.S. sites. One site had groundwater with a pH in the range of 7.4–7.8 and an alkalinity of 120–160 mg $CaCO_3$/L. The other site had surface water with a pH that was manipulated in the range of 7.6–9.5 and alkalinity of 25–60 mg $CaCO_3$/L.

Some significant findings from this study were

- Increasing oxygen (O_2) over the range of 1–27 mg O_2/L did not increase iron release or red water. At the low end of the range of dissolved oxygen levels tested, increasing dissolved oxygen actually decreased iron release. This is contrary to the conventional view that increasing dissolved oxygen increases corrosion; however, the process of iron release is distinctly different from that of classical corrosion.
- At one site it appeared that iron release from unlined cast iron was strongly affected by carbonate, buffer intensity, or a combination of both, within the alkalinity range of 10–60 mg $CaCO_3$/L. Very high alkalinity (>100) had very little impact on iron levels or red water. As TIC increased from 2 to 15 mg C/L at pH levels of 7.6 and 9.0, iron levels dropped significantly. Based on these data, water systems can minimize red water by increasing TIC. This supports the earlier finding on the relationship between buffer intensity and corrosion of unlined cast iron.

Particle Formation by Iron-Pipe Corrosion

A recent literature review summarized the important aspects of iron-pipe corrosion by drinking water (McNeill and Edwards 2001). The review distinguished three distinct, but related, problems. First, pipe mass was lost through oxidation to soluble iron sludge or iron-bearing scale. Second, the scale could accumulate to form thick tuberculation that increased head loss and decreased water capacity. Third, the release of soluble or particulate by-products of iron corrosion into the water decreased its aesthetic quality and often led to consumer complaints of red water at the tap. The conclusion was that iron-pipe corrosion is extremely complicated and is affected by practically every physical, chemical, and biological parameter in water distribution systems. The work included a summary of key factors that utilities must evaluate to mitigate corrosion problems (McNeill and Edwards 2001).

With respect to water quality, only an increase of pH and alkalinity had any beneficial effect on decreasing red-water problems. The presence of microbiological activity, NOM, or free carbon dioxide was detrimental. Measures to control by-product release, such as the addition of phosphate or silicate inhibitors, seemed to have mixed results. Even increasing disinfectant residue had a detrimental effect.

A new topic in the research on the composition of sediment in water mains is the work currently being conducted to develop a fundamental understanding and model of chemically induced iron release (Lytle and Snoeyink 2000). One aspect of the model includes the role of iron colloids and particles following the oxidation of Fe(II). The colloid properties are important in two aspects. First, they could conceivably be incorporated into or deposited on the existing scale. Second, significant concentrations of suspended iron colloids/particles are responsible for red water. The results clearly show that water chemistry parameters such as iron concentration and pH affect the properties of iron particles and iron colloidal suspensions.

Use of Corrosion Inhibitors

Adding corrosion inhibitors is another strategy for preventing the formation of discolored water by cast-iron pipes. This method is generally applied in the United States, France, Germany, and the United Kingdom. Inhibitors frequently used for this purpose include phosphates and silicates (Hambsch and Wagner 1998).

In the United Kingdom, sodium silicate and sodium polyphosphate are both used as water treatment chemicals in an attempt to reduce discolored water from iron-bearing groundwaters and corrosion of unlined cast-iron mains (Williams 1990). Corrosion tests and field trials showed that these chemicals are, in general, successful at reducing discoloration from iron-bearing groundwaters. These chemicals also sometimes lead to improvements in problems associated with corrosion rates and discoloration. The exact circumstances and manner in which these chemicals are effective is not yet clear.

Experiments with a silicate/carbonate combination demonstrated that the release of iron into the water from galvanized mild-steel pipe could be effectively controlled without using phosphate. The protective layers formed by the inhibitor also showed good adhesion and mechanical stability (Schumacher, Wagner, and Wehle 1993).

The effectiveness of adding phosphates to combat discolored water is still under discussion. On the one hand, the study into the physical properties of iron particles in water showed that orthophosphate has the greatest effect on the behavior of iron colloids. The physical changes in colloid properties affected by the presence of orthophosphate resulted in decreasing color and turbidity (Lytle and Snoeyink 2000). On the other hand, the effect of phosphate inhibitors on corrosion of iron pipes under stagnant water conditions showed no effect in two thirds of the experiments conducted. In the remaining experiments, adding orthophosphates and zinc phosphate resulted in an increase of iron concentrations at the tap (McNeill and Edwards 2000).

Recently, tests on the effects of different water quality parameters on iron release (Clement et al. 2002) showed that

- Dosing of orthophosphate (PO_4) can increase the likelihood of red-water problems, especially if pH is altered. In the aforementioned study, orthophosphate was tested in the pH range of 7.4–7.8. Orthophosphate at a residual level of 3 mg/L phosphate (at a pH in the range of 7.3–7.8, and at an alkalinity of 30–35 mg/L $CaCO_3$) reduced iron levels over a 6-month period. Short-term testing in the higher alkalinity water at the University of Illinois did not show any significant effect of orthophosphate on iron release. The lack of any effect in the harder water may result from the short testing time or the possibility that orthophosphate could not penetrate the scales. This warrants an important cautionary note—differences in scale composition may affect the efficacy of orthophosphate in reducing iron levels.

- Sodium silicate has long been recognized as a means of reducing red water by sequestering. Test results showed that sodium silicate in low alkalinity water (30–40 mg/L $CaCO_3$) and pH 8.5–9.0 also reduced the amount of iron from unlined cast-iron pipe.

Controlling Biofilm Formation in the Network

Treated water should have a high degree of chemical and biological stability to prevent undesirable changes in water quality during transportation, storage, and distribution. Growth of microorganisms can lead to consumer complaints about the taste, odor, or discoloration of water (van der Kooij et al. 1999). Water leaving the treatment plant and materials used in drinking water distribution should, therefore, be biostable.

Furthermore, inorganic particles such as hydroxides and oxides of iron and manganese can be trapped in biofilms. Biofilm sloughing can enhance sediment accumulation in peripheral parts of the distribution system. The sediment provides an environment in which certain types of microorganisms (such as *Aeromonas*) and, also, invertebrates can multiply. To understand and prevent these undesirable quality changes, the growth-promoting properties of water and materials in contact with water have been evaluated in The Netherlands (van der Kooij et al. 1999).

Biological Stability of Water

Microbial utilization of biodegradable compounds, which originate from treated water or from exposed material, promotes biofilm development. Assimilated organic carbon and biodegradable organic carbon tests are commonly used to assess the concentration of growth-promoting organic compounds present in water. These tests do not yield information about the rate at which these compounds are used. In addition, the effects of nonorganic growth-promoting compounds (such as ammonia and sulfides) cannot be evaluated using these tests.

The biofilm monitor has been developed to assess the rate and extent of biofilm formation on surfaces exposed to water. In this simple test, ATP analysis is used to measure the accumulation of biofilm material. ATP, an energy-rich compound, is present in all active biomass. ATP accumulation reflects the response of the biofilm to supplies from sources of organic and inorganic energy. ATP analysis also facilitates rapid testing and applicability to all types of biomass, either suspended or attached.

The biofilm monitor allows two characteristic parameters to be determined:

1. The biofilm formation rate (BFR; picogram [pg] ATP/cm^2/day) of the water (i.e., the slope of the linear increase of the biofilm concentration with time)
2. The biofilm formation potential (BFP; pg ATP/cm^2) of water (i.e., the maximum biofilm concentration determined after an exposure period of 150 days)

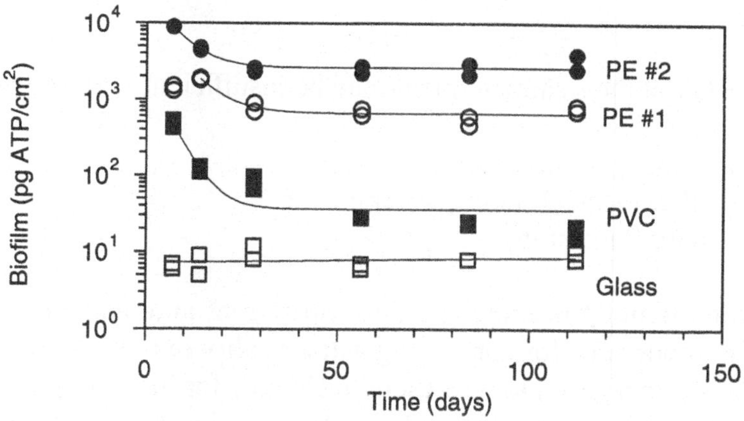

Figure 6.8 Biofilm concentrations on materials in the BFP test. PVC = unplasticized polyvinyl chloride; PE = polyethylene; glass = control material. *(Source:* van der Kooij and Veenendaal [1993]. Reprinted from *Proceedings of 1993 AWWA Water Technology Conference,* by permission. Copyright © 1994, American Water Works Association.)

In many cases, the BFP is not reached within the applied exposure period of 150 days. BFR values in treated water in The Netherlands usually are below 30 pg ATP/(cm^2/day), ranging from less than 1 pg ATP/(cm^2/day) for slow sand filtrates to about 100 pg ATP/(cm^2/day) for drinking water prepared from groundwater that contains ammonia and methane.

Further deposition of iron and manganese in the biofilm monitor can be measured. These parameters, which are expressed as the iron and manganese accumulation rate, respectively, can be useful tools for determining the processes that lead to the generation of discolored water.

Biological Stability of Materials

Certain materials in contact with drinking water can enhance the multiplication of microorganisms. Proper selection of biostable pipe materials is needed to prevent release of organic compounds that promote biofilm formation.

Currently in Europe, three national test procedures are applied to determine the potential microbiological growth of materials. The British Standard BS 6920 measures dissolved oxygen depletion as a surrogate for microbial growth. The DVGW (German Technical and Scientific Association on Gas and Water) W270 method measures the volume of the biofilm that is formed on the surface of materials. In The Netherlands, the biofilm concentrations on materials are measured in the BFP test. This BFP value is defined as the average ATP concentration on the surface of materials in the period between 8 and 16 weeks of incubation at 25°C (van der Kooij and Veenendaal 1993). Figure 6.8 shows biofilm concentrations on materials in the BFP test.

The support of growth of biofilms by corroding cast-iron mains is also a point of interest. Some basic understanding is needed on this subject to gain insight into the generation of discolored water.

CONCLUSIONS

Water quality at the treatment plant may be insufficient with respect to

- Iron and manganese removal
- Biostability of the drinking water
- Corrosivity for cast iron

The quality of and procedures used for operations and maintenance of the treatment plant are of paramount importance for controlling water quality in both production and distribution.

The filtrate scan appears to be a useful technique for increasing the quality of the water that leaves the pumping station. Turbidity and particle counting are especially useful. The techniques used in the filtrate scan do not yield any biological information. Moreover, the filtrate scan does not always result in sufficient information to explain inadequacies in the treatment plant.

Cast-iron pipe corrosion can be restricted by conditioning the drinking water. The pH and the TIC content of the water especially need attention.

The effectiveness of adding phosphates to combat discolored water is still under discussion. Silicate addition decreases the iron release by cast iron.

To avoid biofilm formation in the network, the use of biostable water as well biostable pipe materials must be considered.

An integral strategy is needed to simultaneously deal with as many processes that threaten quality as possible. A cost-benefit analysis should determine which control measures ought to be applied in which circumstances.

CHAPTER 7
STEP 2: PREVENT SUSPENDED SOLIDS FROM SETTLING

INTRODUCTION

Discolored water is related to velocity—at low velocities, particles that cause discolored water can settle out or form and settle out. These particles can be resuspended with sudden change in flow conditions. However, adverse flow conditions do not cause discolored water; they only promote sediment formation and resuspension.

Restricting the number of particles in water main systems can help to prevent the formation of discolored water. Another possibility is to ensure that particles cannot settle in water main systems. To do this, the design of the network can be optimized by avoiding long stagnation times and dead ends in the network.

EFFECTS OF HYDRAULIC PROCESSES

Determining the effects of hydraulic processes on the generation of discolored water is a complex process that requires insight into the hydraulic conditions and characteristics of the sediment. The disturbance of any sediment present in water mains, which is the most common cause of discolored water, can happen when the direction or velocity of the water changes. Changes in velocity can occur for a number of reasons, including a water main burst, opening or closing of valves, bringing mains back into use after repair, and using water for fire fighting.

Stagnation time has an influence on the occurrence of discolored water. The most frequently cited reason for discolored water is the presence of dead-end mains. However, dead ends can be seen as a specific situation—because of the absence of flow, particulate matter can easily settle; when the water flows, this material will be resuspended.

Thames Water in the United Kingdom designed a test facility to determine the behavior of loose deposits and of biofilm in a "perfect" water distribution system. A 1,400-m long, 102-mm-diameter pipe was used for these experiments. It was clear from experiments using this test facility that the majority of suspended particles entering this pipe passed through apparently unchanged. However, loose deposits were shown to accumulate over time at certain points. These loose deposits were readily lifted into suspension when velocities were significantly increased (Woodward, Colbourne, and Holt 1995).

Cast-iron pipe sections from the distribution system were installed in this testing facility and the system was operated with an alternate stagnation and flow regime (Smith et al. 1999). Preliminary results showed that the facility was capable of generating corrosion deposits similar to those from the network and of simulating discolored water events. It also provided time-sequence data on the production of discolored water in stagnant-flow conditions.

Continuous monitoring of turbidity over the stagnation and flow cycle in this experimental pipeline showed that most turbidity generation occurred during stagnation. However, periods of flow were associated with mainly nonturbid water, with distinguished but very short turbidity peaks occurring at the start of the flow cycle. The detailed tracking of the turbidity and flow regime demonstrated the significance of the stagnation periods for iron release during intermittent flow regimes (Smith et al. 2000).

It has also been demonstrated in the field that, in a corroding cast-iron main, turbidity increases during stagnation and decreases as the velocity increases. This can be seen in patterns found by continuously monitoring turbidity in cast-iron mains (van den Hoven et al. 1994). In the case of rusting cast-iron pipes, a specific pattern of iron release is observed. The pattern of increasing turbidity between 11:00 p.m. and 7:00 a.m., with an abrupt decrease in turbidity when the water is first used in the morning, is caused by the absorption of corrosion products by the water when it is stationary. When the water is first used in the morning, turbidity decreases abruptly because of the replenishment of the pipeline's contents (Figure 4.18).

Optimizing Water Main Designs

Another possibility is to ensure that the dimensions of water main systems prevent sediment from settling. These water main systems are known as self-cleaning systems. In 1999, two new guidelines were published on designing self-cleaning distribution networks in The Netherlands (van den Boomen and Vreeburg 1999).

Because of its branched construction and relatively small pipe diameters, a self-cleaning distribution network has a unidirectional flow and higher velocity than a conventional network.

Conventional Distribution Networks

In The Netherlands, the structure and capacity of conventional distribution networks are determined by conventional requirements for water for fire fighting. The merging, since 1930, of supplies for drinking water with those for fire fighting has resulted in conventional distribution networks that consist of pipes with diameters of 100 mm or more, arranged in a ring structure. These pipes meet the fire department's conventional requirements for fire fighting. Figure 7.1 shows an example of a conventional distribution network. The conventional requirement for water for fire flows is much higher than the normal requirement for drinking water. The large diameters result in low velocities and in high residence times in the pipes during periods of normal drinking water consumption. The distribution network's ring structure causes the water to have alternating flow directions. The low velocities, combined with alternating flow directions, result in sediment accumulation in the pipes. Occasional higher demand, for example when a fire hydrant is used, causes sediment to be resuspended and carried along. Customers then see discolored water at the tap. Under certain conditions, the long residence times can also result in deterioration in bacteriological conditions.

Characteristics of conventional distribution networks are

- Fire flow requirements that dominate drinking water requirements
- Large diameters; fire hydrants of 100 mm or larger
- Low velocities during periods of normal drinking water consumption
- Alternating flow directions
- Long residence times; high residence time spread
- Accumulation of sediment
- Discolored water complaints

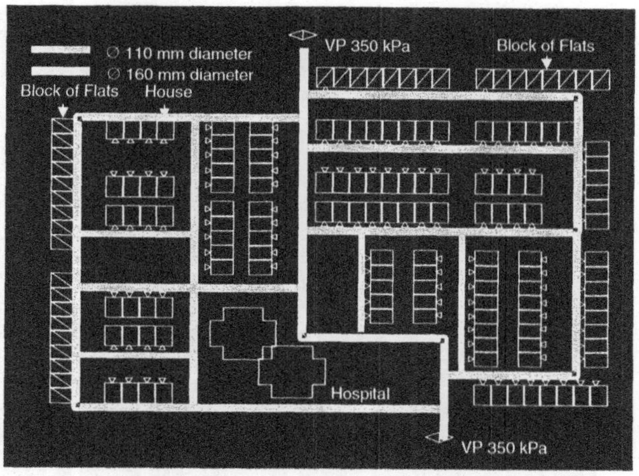

Figure 7.1 Conventional distribution network

Self-cleaning Distribution Networks

Water companies in The Netherlands have been using the new design guidelines for distribution networks since 1999. The starting point in the guidelines is that sediment does not remain in pipes for long periods if the water achieves a velocity of at least 0.4 m/sec once a day. This starting point is based on research with fine sediment in a smooth tube, carried out in a laboratory rig. The number corresponds reasonably well with the related critical (wall) shear stress for the start of resuspension of noncohesive material according to the familiar Shields diagram (Shields 1936). Figure 7.2 shows an example of a self-cleaning distribution network. The white pipe is part of the meshed principal distribution network. The ring structure, which guarantees a supply, is the feed for the self-cleaning, branched distribution sections.

Characteristics of self-cleaning distribution networks are

- Designed for normal drinking water requirements
- A velocity of 0.4 m/sec once a day in the end pipes
- A unidirectional flow
- Short residence times
- No accumulation of sediment
- Possibility for incorporating adequate water supplies for fire fighting

The branched distribution sections may vary in size. A section may sometimes consist of just one street; at other times, there may be several streets. The branched distribution sections are designed for a unidirectional flow and a sufficient velocity for momentary domestic demand. Because of this, there is no opportunity for sediment to settle, and residence times are short.

From a technical design point of view, this means that pipes can be made increasingly thinner as fewer homes are connected to them. Moreover, it has proved possible, in consultation with the fire department, to incorporate water supplies for fire fighting that are in keeping with modern developmental requirements.

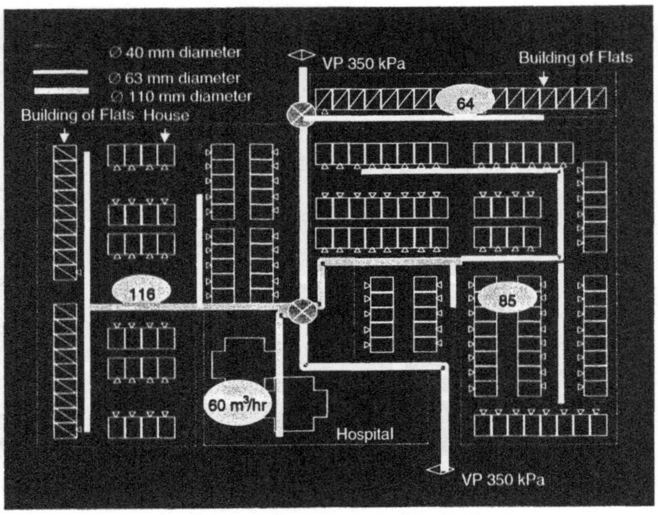

Figure 7.2 The principle of a self-cleaning distribution network

In The Netherlands, the Fire Brigade Association has developed primary capacity guidelines for fire fighting with water:

- For houses built before 1991, 60 m^3/hr. These houses are not constructed according to the Building Resolution (1991).
- For houses built after 1991 (without additional risk factors), 30 m^3/hr. Those houses are constructed to conform to the Building Resolution (1991), which means that they have a resistance to fire transmission to the adjacent building of at least 1 hour. Risk factors that require more than 30 m^3/hr capacity for fire fighting with water are more than 25 persons present at the same time, more than four storage buildings on the premises, and onsite storage of chemicals or other dangerous substances, among others.

Design Steps for a Self-cleaning Distribution Network

The design process is divided into three steps:

1. Design a branched self-cleaning drinking water network based on the maximum momentary demand for drinking water.
2. Incorporate the supply for fire fighting into the drinking water network.
3. Check the design for pressure losses:
   - During drinking water consumption
   - During use of fire hydrants

Various methods are currently used to determine the maximum momentary demand for drinking water that the distribution network must be designed to meet. The $q\sqrt{n}$ method is the most widely used in The Netherlands. This method is used to calculate the maximum volume flow that occurs in a drinking water supply system with various tap points. The $q\sqrt{n}$ method assumes that the likelihood of volume flows occurring simultaneously at tap points falls as the number of

Table 7.1
Tap units for various tap points (for an average home)

Tap points per home	Number of tap units, TU [–]
Toilet 1	0.25
Toilet 2	0.25
Hand basin tap 1	0.25
Hand basin tap 2	0.25
Kitchen mixer tap	4
Dishwasher	4
Bath/shower mixer tap	4
Washbasin tap	1
Washing machine	4
Garden tap	4
Total	22

tap points connected to the drinking water supply system increases. After all, not all the taps will be operated at once. The $q\sqrt{n}$ method is used to determine the maximum volume flow for housing, offices, and shopping centers:

$$q_m = 0.083 \cdot 10^{-3} \cdot \sqrt{\Sigma \mathrm{TU}} \tag{7.1}$$

The tap unit (TU) is a calculation unit for determining the maximum momentary volume flow at tap points and appliances. One TU is equivalent to a flow volume of $0.083 \cdot 10^{-3}$ m^3/sec. The number of tap units in a building (e.g., an average home) is calculated using a table similar to Table 7.1.

Dutch water companies recommend that the garden tap should no longer be included in the calculation because use is irregular. The total number of tap units for a medium-sized home, then, is 18.

Using the new design method, a residential district or a pipe section is seen as one large domestic installation in which each tap in each home is a separate tap point. The total number of tap units equals the number of homes multiplied by the number of tap units per home (TU_h).

$$\Sigma \mathrm{DU} = n \cdot \mathrm{TU}_h \tag{7.2}$$

The momentary consumption of several homes is thus calculated using Equation 7.3, in which a tap unit is equivalent to $0.083 \cdot$ L/sec $= 0.083 \cdot 10^{-3}$ m^3/sec.

$$q_{m,n} = 0.083 \cdot 10^{-3} \cdot \sqrt{n} \cdot \sqrt{\mathrm{TU}_h} \tag{7.3}$$

The number of tap units, TU, depends on the type of home and varies between 12 and 22. The graph in Figure 7.3 shows the momentary consumption for three types of homes, calculated using the $q\sqrt{n}$ method.

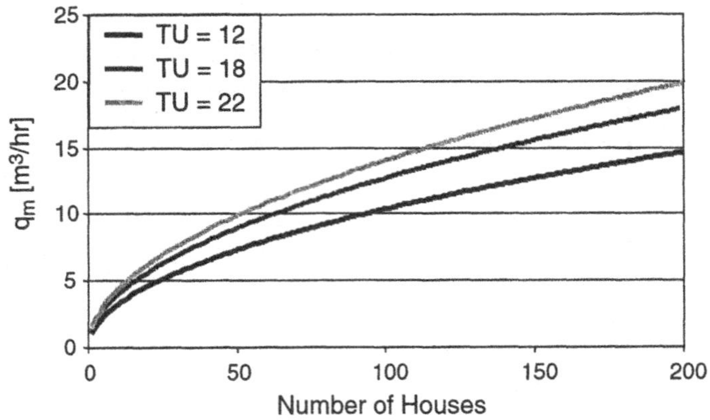

Figure 7.3 Momentary consumption according to the $q\sqrt{n}$ method shown for three homes

Risk assessments on the newly designed networks have been carried out with respect to delivery capacity and pressure losses during sprinkling. Results showed that in a branch designed on self-cleaning capacity:

- The demand that could be supplied was at least twice that of the designed demand.
- At least 50% of the home owners could simultaneously sprinkle their gardens without unacceptable pressure losses.

Fluctuating demand can be addressed by designing on normal demand and by carrying out a risk assessment and sensitivity analysis on fluctuating or increasing future demand. The sensitivity for fluctuating demands in The Netherlands was not that great. The systems provide more capacity than one would expect.

VALIDATION OF THE NEW DESIGN GUIDELINES

Starting Points

As mentioned previously, a starting point of the design guidelines for a self-cleaning distribution network is that sediment will not remain in the pipes for a long time, if water achieves a velocity of at least 0.4 m/sec once a day. This design velocity was determined for the guidelines on the basis of experience gained from operations at water companies. Volume flow and turbidity measurements after construction have demonstrated that sediment does actually move through the pipes.

To further implement the design guidelines, it was necessary to obtain more information on the behavior of sediment and to validate the design velocity. The investigation consisted of a literature study of sediment behavior in distribution mains and analyses of sediment behavior in tubes set up in a laboratory (Beuken 2001, Beuken and van Lieverloo 2001).

Composition of Sediment in the Water Main System

In the water main systems, sediment can occur in the following forms:

- As suspended solids in flowing water
- Attached to the pipe wall (e.g., as biofilm or a corrosion layer)
- As loose deposits (moving or otherwise) at the bottom of the pipe

Sediment in distribution mains mainly comes from

- The treatment plant (e.g., because particulate matter is not completely removed, because filters are put into operation too soon, or because chemical products are added)
- Pipe material (e.g., because of corrosion of cast-iron pipes or because of lining coming loose)
- Contamination entering the pipes (e.g., because of repair work, fractures, or inexpert use of fire hydrants)

Chemical, biological, and hydraulic causes result in various secondary processes, such as the erosion of corroded cast iron, ongoing coagulation of colloidal particles to form particles that can settle, the conversion of waterborne organic matter into a biofilm, and the growth of invertebrates.

As discussed in chapter 2, the largest component by volume of the sediment consists of small particles (<50 μm) (van Lieverloo et al. 2002). Most sediment of >500 μm is inorganic and is largely accounted for by corrosion particles. More than half of the 100- to 500-μm fraction consists of organic matter (abraded biofilm and isopod droppings). Minerals account for about 15% of the total volume of sediment. In the case of smaller particles, we can assume that the percentage of organic matter is higher.

Measurements Made Using a Test Rig

A laboratory test rig, shown in Figure 7.4, was designed to study sediment transport. It consisted of a transparent Perspex pipe, through which water flowed and into which sediment was introduced. The volume flow could be precisely set using a membrane valve and a volume flowmeter. The water's velocity was slowly increased in steps while a camera filmed sediment behavior. The measurements were made in PVC pipes of five different diameters (40, 50, 63, 75, and 110 mm).

Tests were conducted using five types of standardized sediments. Standardized sediments were used because they have a uniform composition, their grain size and specific gravity can be accurately determined, and the sediment can be reproduced for subsequent tests.

The standardized sediment consisted of

1. Sand, grain size 180–250 μm, density 2,650 kg/m^3
2. Iron oxide, grain size 45–90 μm, density 3,140 kg/m^3
3. Iron oxide, grain size 180–250 μm, density 3,140 kg/m^3
4. Flour, grain size 63–90 μm, density 1,320 kg/m^3
5. Flour, grain size 150–180 μm, density 1,320 kg/m^3

Figure 7.4 Laboratory test rig for sediment tests

Observations

An analysis of video images recorded the moment at which the initial movement of a few particles became visible and the moment at which all the particles were in motion. The measurements did not reveal any clear relationship between pipe diameter and velocity in terms of the moment at which particles started to move (Figure 7.5). Sand (grain sizes 180–250 μm) first began to move at a velocity of approximately 0.15 m/sec, and all the particles were in motion at approximately 0.25 m/sec. Initial motion in both fractions of iron oxide began at 0.13 m/sec, and all particles were in motion at approximately 0.22 m/sec. The 63–90-μm grain size of flour first began to move at 0.08 m/sec, and all the particles were in motion at 0.15 m/sec. Initial motion of the 150–180-μm grain size of flour began at around 0.10 m/sec, and all particles were in motion at around 0.16 m/sec.

During the measurements, it became clear that the type of sediment on the bottom of the pipe affected the behavior of the sediment. An individual particle began to move sooner than a small heap of sediment. Particularly for sand, and to a lesser degree for the 180- to 250-μm iron oxide fraction, it was clear that the sediment grouped into "islands." The islands took on the shape that offered least resistance to the flowing water. This means that if sediment groups into islands, a higher velocity is required to cause motion. This explains the relatively high velocity that is needed to cause sand to move.

Sediment behavior was filmed from the side in a 63-mm-diameter pipe. This made the distribution of the resuspended sediment visible across the height of the pipe. It appeared that, at a velocity of 0.35 m/sec, all types of sediment roll or float along the bottom of the pipe. This means that the movement of sediment through pipes will be impeded at places where height differences occur in the bottom of the pipe; for example, at the location of reducing fittings. The tap-in points

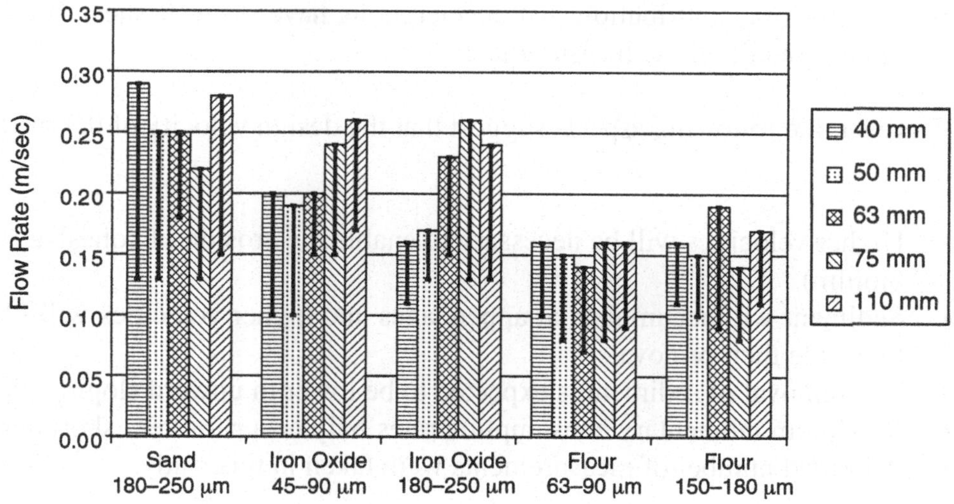

Figure 7.5 Velocity at which initial motion began (bottom of line) and at which all particles were in motion (top of line) for various pipe diameters and particle sizes (μm)

of junction pipes also require further attention. It can be expected that removing fine sediment, if it is at the bottom of the pipe, will be more difficult.

In situations in which the sediment could be readily observed, a comparison was made between the rate at which sediment moved along the pipe and the velocity of the water. This showed that pipe diameter has no effect on the velocity at which sediment starts to move, but that the rate at which the sediment moves along the pipe is higher in pipes of a smaller diameter. Roughly speaking, we can conclude that motion in heavier material (sand and iron oxide) starts at 0.25 m/sec, and, at a velocity of 0.4 m/sec, the sediment moves along the pipe at a rate of 10 m/hr in a 110-mm-diameter pipe, 18 m/hr in a 75-mm-diameter pipe, and 30 m/hr in a 63-mm-diameter pipe. Motion in lighter material starts at 0.10–0.15 m/sec and, at a water velocity of 0.2 m/sec, the speed is of a much higher order of magnitude (250 m/hr). This shows that lighter sediment will move through the water main system more readily than heavier sediment.

Discussion

A number of measurements were made to validate one of the starting points for designing self-cleaning networks (that no sediment must accumulate if a velocity of 0.4 m/sec occurs once a day). From the limited number of measurements, it appears that motion starts in practically all settled sediment particles at a velocity of approximately 0.25 m/sec. Because this result was seen in a test situation, various aspects must be considered before this is applied to a field situation.

The following aspects mean that consideration can be given to the possibility of lowering the self-cleaning design velocity of 0.4 m/sec:

- The tests showed that sediment was in motion at lower velocities.
- Most of the sediment is light and organic, meaning that it begins to move at a lower velocity than heavier sediment.

- The positive contribution that accelerations have on wall shear stress was not taken into account in these measurements.

However, the following aspects suggest that the design velocity of 0.4 m/sec should not be lowered:

- Higher velocities will be necessary to enable the erosion of cohesive particles (such as biofilm).
- Sediment will accumulate at appendages and higher velocities will likely be required to enable their removal.
- The removal of sediment is expected to be lower in upward sloping pipes.
- Maximum momentary consumption rates only occur for very short periods.
- A limited number of measurements were taken in this study.

In view of these factors, the design velocity of 0.4 m/sec has been maintained for the time being. Further studies should indicate whether a change in the design velocity is necessary.

EFFECT OF CONSUMPTION PATTERNS

Research into the background of self-cleaning capacity shows that pipes can also be self cleaning when velocities are low. The self cleaning is attributed to the changes in velocity caused by variations in consumption. Up to now, it had been assumed that sediment could only be resuspended and carried along above a certain high velocity.

The guidelines for designing a self-cleaning capacity assume self cleaning for designs based on a peak velocity of 0.4 m/sec. This value is based on exploratory studies carried out in a laboratory rig, with fine sediment in a smooth tube. The value corresponds reasonably well with the related critical (wall) shear stress for the start of resuspension of noncohesive material according to the familiar Shields (1936) diagram.

The First Consumption-pattern Measurements in the Field

The first field measurements in “self-cleaning end pipes” were made by Brabant Water, the Provincial Water Company of North Holland, and Nuon Water. The measurements showed that, generally, the peak velocity of 0.4 m/sec is not achieved. Figure 7.6 shows an example of the measurements.

The actual peak velocities that occur in end pipes are around 0.1 to 0.3 m/sec, for two reasons:

1. The peak consumption figure used in the design formula is too high.
2. The pipe diameters used are still too large (in connection with the fire department’s water requirements for fire fighting).

These findings resulted in further research, conducted jointly by Kiwa and the Delft University of Technology (Delft, The Netherlands), into the self-cleaning abilities of drinking water pipes (van den Boomen and van Mazijk 2002). The focus of this research was the extent to

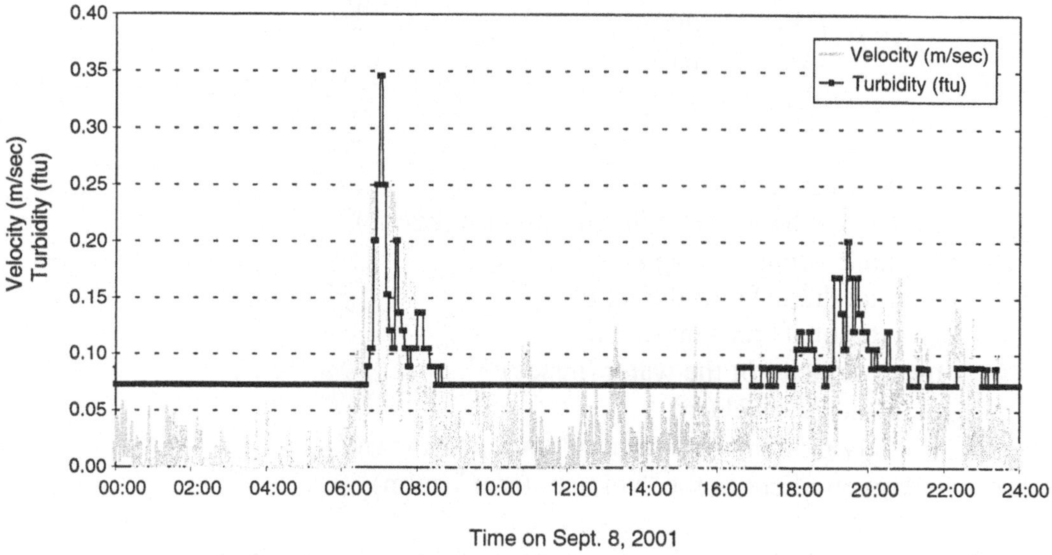

Figure 7.6 Turbidity and flow measurements in a branch of a self-cleaning network, PVC 75-mm diameter, downstream demand of 46 households

which distribution mains are self cleaning when the peak velocity of 0.4 m/sec is not achieved during peak consumption.

Research Into the Effect of Flow Dynamics on Self-cleaning Capacity

This research investigated flow dynamics in a pipe. The effect of a change in velocity on a pipe's self-cleaning capacity was examined, as was the effect of the actual velocity. Because no precise information was available on the (wall) shear stress figure at which cohesive sediment in drinking water pipes is resuspended (thereby leading to self cleaning), the following hypothesis was formulated.

Hypothesis

Because fieldwork and exploratory studies have shown that good results are achieved with designs based on a specified velocity, the wall shear stress figure relating to that velocity was worked out theoretically. The theoretically determined figure for *wall shear stress* is now defined as the *critical shear stress* for the start of resuspension and is used as a design criterion. Once this critical value is adopted, theoretical considerations can be used as a basis for determining the effects of velocity, changes in velocity, and the diameter of the pipe on the critical shear stress.

Hydraulic Conditions and Wall Shear Stress

The wall shear stress that occurs in a pipe can be worked out from the general motion comparison for flowing water under pressure in a closed pipe, and is described as follows:

$$\tau = -\rho \cdot R \cdot \left(\frac{\partial u}{\partial t} + g \cdot \frac{\partial \varphi}{\partial x}\right) \tag{7.4}$$

where:

τ = shear stress (N/m^2)
ρ = density of the water/sediment mixture (kg/m^3)
R = hydraulic radius (= ¼ · D; m)
D = internal diameter of the pipe (m)
g = gravitational constant (m/sec^2)
u = mean velocity of the water (m/sec)
φ = piezometric level (m)
$\partial u/\partial t$ = velocity gradient or "the acceleration term" (m/sec^2)
$\partial \varphi/\partial x$ = energy gradient or the "velocity term" (m/m)

The comparison shows that shear stress depends on the diameter of the pipe, the velocity term ($\partial \varphi/\partial x$), and the acceleration ($\partial u/\partial t$) of the water.

Relationship Among Velocity, Velocity Change, and Diameter

Again, the starting point is that the critical shear stress for self-cleaning capacity in drinking water distribution mains occurs at a constant velocity of 0.4 m/sec in a 110-mm-diameter PVC pipe. At this value, the theoretical figure for wall shear stress is 0.47 N/m^2. Given this value for wall shear stress, it is now possible to determine—for any pipe diameter—which combination of velocity and velocity change produces this value of wall shear stress (Figure 7.7). Figure 7.7 shows that a 110-mm-diameter PVC pipe meets the basic requirement for self cleaning at a constant velocity of, for example, 0.10 m/sec and a velocity change of 0.0166 m/sec^2. This means that, if the water velocity in this pipe changes from, for example, 0.10 m/sec to 0.12 m/sec in 1 second, the wall shear stress that occurs is the same as that for a constant velocity of 0.4 m/sec (the starting point for self cleaning). It is very likely that such velocity changes occur in the field.

A cautious conclusion, therefore, is that in end pipes in which the peak velocity of 0.4 m/sec is not achieved, sufficient shear stress nevertheless occurs to keep the pipe clean. In other words, at lower velocities, the velocity change makes a greater contribution to the self-cleaning capacity of a pipe than does the actual velocity. This means that velocity changes in water cannot be neglected when designing self-cleaning distribution mains.

The diameter of the pipe is another factor that affects the wall shear stress occurring in a pipe as a result of flow conditions. Figure 7.7 shows the effect of the velocity and velocity change on wall shear stress for different pipe diameters. For example, it shows that a stable velocity of 0.34 m/sec in a 40-mm PVC pipe produces the same shear stress as a stable velocity of 0.4 m/sec in a 110-mm-diameter PVC pipe.

CONCLUSIONS AND RECOMMENDATIONS

Current design guidelines for designing drinking water pipes with a self-cleaning capacity are based on a stable peak velocity of 0.4 m/sec. The design guidelines appear to meet requirements in the field and are on the safe side.

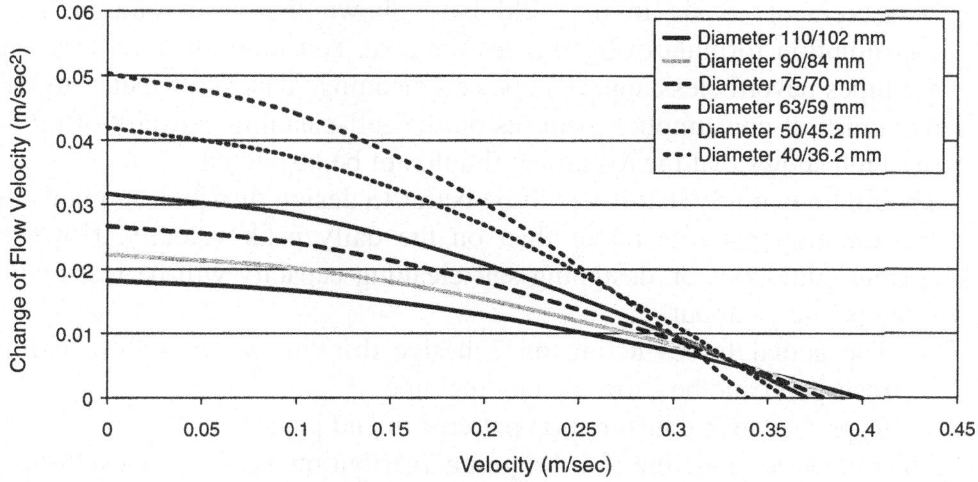

Figure 7.7 Combinations of flow velocities and changes of flow velocities that reach a shear stress of 0.47 N/m². Valid for PVC drinking water pipes with a wall roughness of 0.05 mm (for various PVC pipe diameters).

The design guidelines could be improved by incorporating designs based on the theoretically determined wall shear stress that occurs at this stable peak velocity of 0.4 m/sec in a 110-mm-diameter PVC pipe.

This would make it possible to design for lower stable velocities, provided the velocity changes caused by consumption were sufficient to achieve this theoretically determined wall shear stress. A calculation instrument is being developed for this calculation. Improving the guidelines would provide greater design scope for meeting the self-cleaning requirement.

Conclusions

Based on this work, we can draw the following conclusions:

- The results of this research have made it possible to
  - Determine the wall shear stress that occurs in existing pipes and for a known consumption pattern. This gives an indication of the size of the self-cleaning capacity in those pipes.
  - Design water main systems for new constructions and a known consumption pattern, based on the required theoretical wall shear stress
- The new design guidelines for designing distribution networks with a self-cleaning capacity appear to meet requirements in the field and are on the safe side. The design guidelines are
  - Design based on a daily peak velocity of 0.4 m/sec
  - Ensure a definite flow direction or a branched network structure
- Applying the new design guidelines for distribution networks results in better water quality, less cleaning work, and a saving in construction costs that can be as much as 30% of the costs of the materials used.

- Measurements made in the field have shown that even though the current peak consumption formula overestimates the peak consumption level that actually occurs, the pipes nevertheless appear to be self cleaning. This is attributed to the effect of the dynamics of consumption patterns on the self-cleaning capacity of pipes. Theoretical research shows that the dynamics should not be neglected.
- The findings imply that it would be better to design distribution mains on the basis of consumption patterns rather than on the daily peak velocity. However, improving current guidelines on designing self-cleaning capacity will be worthwhile only when more is known about
  - The actual forces acting on cohesive drinking water sediment that cause it to resuspend and be carried along the pipe
  - Representative consumption patterns in end pipes
- The improved guidelines for designing distribution mains with a self-cleaning capacity will also have to be scaled up for the main pipe and carrier pipe level.

This research has revealed that the velocity *change* of water in distribution mains makes a major contribution to the self-cleaning capacity of the pipes. It is therefore important to know what the consumption patterns are at different points in distribution networks. This means more consumption-pattern measurements will have to be made. Consumption patterns are highly influenced by the size of background consumption and the degree of simultaneity in that background consumption. The wall shear stress in a pipe is not equal to the shear stress on the cohesive sediment.

Recommendations

Research is recommended into the forces that shear stress exerts on cohesive sediment in the distribution mains. This research should finally result in the capability to model sediment transport in drinking water pipes. A sediment transport model for drinking water pipes could form the basis for developing models of fouling in water main systems, as well as calculation models for designing self-cleaning water main systems.

CHAPTER 8
STEP 3: REMOVE SEDIMENT FROM MAINS EFFECTIVELY AND IN DUE TIME

WATER MAIN CLEANING

Mains are cleaned to remove sediment and slime that accumulates on the walls of pipes in distribution networks. The cleaning effect is achieved because a high velocity is generated along the pipe wall, producing a wall shear stress sufficient to loosen sediment and biofilm from the pipe wall. The flowing water then carries these along. In The Netherlands, various methods are used to clean mains:

- Flushing with water
- Flushing with water and air
- Swabbing (using pigs)

When flushing with water, opening a fire hydrant or a flushing point installed for that purpose produces a high water velocity. When flushing with water and air, introducing compressed air into the pipe at the same time as increased water velocity increases the cleaning effect. In the case of swabbing, a plastic pig is transported through the pipe. Because the pig has a slightly larger diameter than the pipe's internal diameter, the sediment is scraped from the wall and pushed along in front of the pig.

STARTING POINTS FOR FLUSHING

The flushing velocity must be sufficient to resuspend and carry sediment along the pipe. Practical experience shows that a velocity of 1.5 m/sec effectively removes sediment. Flushing at a lower velocity may loosen sediment but may not carry it along the pipe, and this may even result in an increase in the number of complaints when water delivery resumes.

When flushing, some of the sediment will move more slowly than the mean velocity of the water. This means that the total flushing volume must be higher than the content of the pipe. Practical experience has shown that, at a flushing volume of 3 to 4 times the content of the pipe, the turbidity of the flushing water is more or less equal to that of the incoming water. However, during flushing, it is necessary to verify that the pipe has been sufficiently flushed clean. The easiest method is to use a white bucket to visually assess the water. Flushing should be stopped a few minutes after it has been established that the water is clear. The verification can also be made using a portable turbidity meter.

In 1996, Kiwa conducted a comparative study of the effects of flushing with water, water–air flushing, and swabbing (Vreeburg 1996). The comparison revealed that if a water flushing operation is carried out properly, the higher cleaning effect of swabbing and water–air flushing does not compensate for the benefits gained because flushing with water is easier to perform.

Practical experience has shown that flushing with water involves a number of limitations or disadvantages, such as

- Pipes made of corroded cast iron are difficult, if not impossible, to scour because corrosion-covering layers may be damaged or because the necessary velocities cannot be achieved.
- There may be limitations on the pipe's water carrying capacity.
- The capacity for flushing water discharge may be limited.

The specific disadvantages of water–air flushing were that

- Introducing air raises the risk of contamination, and extra filters must be fitted to prevent oil from entering the water through the airflow.
- The cleaning effect of water–air flushing is limited to diameters smaller than 250 mm.
- The extra turbulence means that a great deal of caution is necessary in corroded cast-iron pipes. Because corrosion-covering layers may be loosened, there is a risk of accelerating the corrosion process.
- The air still present in the pipe must be removed after water–air flushing. Because this often proves difficult in practice, intensive subsequent flushing with water still frequently turns out to be necessary.
- In practice, the cleaning effect on longer pipes is reduced because the water and air in the pipe separate. Therefore, after a certain length, air and water flow alternately through the pipe, which reduces the cleaning effect.

The specific disadvantages of swabbing were that

- Swabbing introduces an extra risk of contamination, both from introducing the pigs and because pigs wear out, leaving pig material remains in the water main system.
- Swabbing cannot be used in sections with corroded cast-iron pipes.
- The presence of butterfly valves may limit the length that can be swabbed.
- Swabbing is relatively expensive, partly because of the relatively major modifications required for installing the launching and receiver fittings and because of the relatively rapid wear rate. The study assumed that six to seven pigs are used for each pipe section of 2.5 to 4 km.
- Subsequent flushing is still necessary.

Three flushing approaches are common in the United States (Antoun, Dyksen, and Hiltebrand 1999; Kirmeyer et al. 2000):

1. Unidirectional flushing. This involves isolating pipe segments, typically by closing appropriate valves and operating the hydrants from source to periphery in an organized, sequential manner. It can be used for system-wide flushing to prevent water quality deterioration (on an on-the-spot basis) or in response to a specific water quality concern. System-wide unidirectional flushing, in particular, requires careful preparation and supervision during implementation. Unidirectional flushing is most closely related to the flushing practices that are recommended in this study.
2. Conventional flushing. This involves opening hydrants in a specific area of the distribution system, until preselected water quality criteria are met. Because valve isolation

is not practiced, flushing velocities are not maximized—as water to the hydrant flows from several mains and the resulting velocity in each individual main remains low.

3. Continuous blow-off or bleeding. This is used at dead ends with severe high-retention problems, whereby a constant level of blow-off water is extracted from the network to maintain residual disinfectant levels or reduce water age.

THE STANDARD CLEANING PROGRAM

Flushing is most effective when carried out using what is known as a clean-water front. Before starting the cleaning program, the clean sections must be defined. The section may be the pumping station or a large trunk main. Measurement is generally required to determine if a trunk main can be used as a clean-water front. The clean-water front follows the flow direction of the network. Working according to this concept maximizes assurances that recently cleaned areas will not be refouled by hydraulic disturbances in the network.

Before the clean-water front can be defined, information on the pipelines must be collected. Next, the sequence of distribution areas to be cleaned is determined. Finally, calculations are made to determine where flushing points can be located. These points must be selected so that they can be easily and safely accessed and so that there are adequate drainage possibilities.

Although a flushing program can be worked out using maps and hydraulic calculations, this is extremely labor intensive. If the pipe details have been recorded digitally, a computer can do much of this work. Kiwa has developed a computer model, called Flushplanner, for this. The model's output consists of a flushing form, on which the valve adjustments are printed for each pipe being flushed. The form also serves as a means of registering the operations that the flushing team will carry out.

A great deal of information on the main system is generated when a systematic cleaning plan is being drawn up and executed. This can be used to manage the system systematically. Every cleaning program will yield new information on the quality of the water main system, which relates to matters such as the ease of shut-off valve operation, as well as the degree of corrosion and biofilm formation. The information can be used, for example, to establish priorities in the renovation of cast iron, to restyle diameter transitions, and to adequately monitor shut-off valves and fire hydrants.

Working together, Kiwa and the Dutch water companies have drawn up some basic principles for implementing flushing programs:

- Water flushing is the most effective technique.
- When flushing, the flushing rate of the water must be at least 1.5 m/sec.
- The volume of flushing water must be at least three times the volume of water in the pipe until a visual inspection shows that the water is clean.
- The work is performed on the basis of a clean-water front.
- Taking measurements, in addition to carrying out the cleaning activities, increases effectiveness.

In addition to these technical requirements, attention must be paid to ensuring that the cleaning program will

- Be well organized (and that the activities and relevant observations will be properly recorded while the work is being performed)
- Cover all pipelines in the area to be flushed
- Be executed efficiently and based on realistic time planning
- Be preceded by a check of the functioning, accessibility, and correct numbering of the shut-off valves

Special Attention for Cast Iron

Cast-iron pipelines of a smaller diameter (especially smaller than 150 mm) require special attention. Hydraulic capacity may be lower than the theoretically calculated capacity because of internal corrosion. However, the reduced diameter can mean that the velocities occurring in the pipeline are higher than the rate calculated on the basis of the outgoing velocity and the original diameter. Flushing these pipelines involves a risk of damaging protective corrosion layers. Any such damage will activate the underlying corrosion layers and release iron salts into the water, resulting in an increase in turbidity.

Once the protective corrosion layers are damaged, turbidity should be monitored at regular intervals. An increase in turbidity can be monitored especially well at night when velocities are low (see Figure 4.18). If the former situation is not restored, the pipe section can be expected to experience future turbidity problems. Rehabilitation is recommended in this case.

NEW FLUSHING METHODS

Pulsating Flushing

The cleaning principle in pulsating flushing is to loosen and resuspend sediment particles in the pipe by accelerating the water's velocity. The flow is increased within a short period according to a fixed pulsating pattern. The increased velocity results in a brief but sharp increase in wall shear stress. The acceleration is achieved by the controlled, repeated opening and closing of a valve at the end of the section being flushed (Figure 8.1). In between the periods of increased velocity, flushing takes place at a constant basic velocity, which is sufficient to carry along the loosened particles. Figure 8.2 illustrates an example of a pulsation pattern.

Practical experience has shown that a constant flushing rate of 1.5 m/sec is sufficient to remove sediment in a 400-mm-diameter PVC pipe. The wall shear stress that occurs under those conditions is 3.8 N/m^2.

Assuming that the wall shear stress is the main factor for sediment removal, a wall shear stress of 3.8 N/m^2 would be sufficient for cleaning the pipes. The velocities and accelerations given in Table 8.1 can be used as a guideline for sediment removal.

A test was carried out to test the theory of pulsating flushing in the field (Schaap and van den Boomen 2001). Numerous pulses were introduced in accordance with a predetermined regime. The test was carried out on a 160-mm-diameter PVC pipe of 1,400 m in length. The assumed starting point was that the shear stress produced had to be at least 3.8 N/m^2 and that the content of the pipe section had to be replenished three times. During tests in the field, the flushing volume on the hydrant was digitally measured per pulse and recorded in writing.

A basic velocity of 0.4 m/sec was chosen for the test in the field and a final velocity of 1.0 m/sec was chosen, which had to be reached within a time interval of approximately 6 seconds.

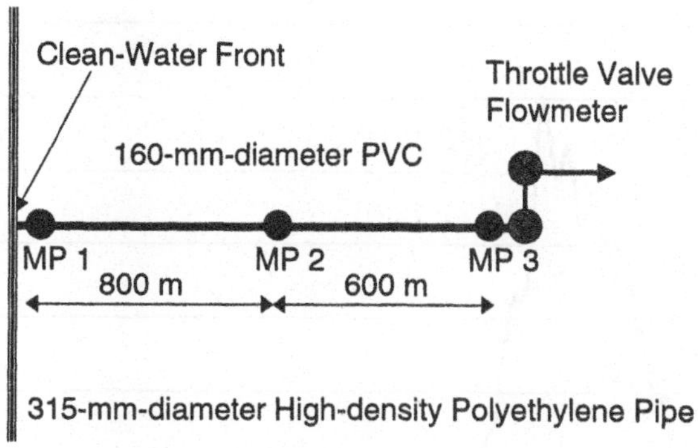

Figure 8.1 Overview of the flushing test and the position of MPs 1, 2, and 3

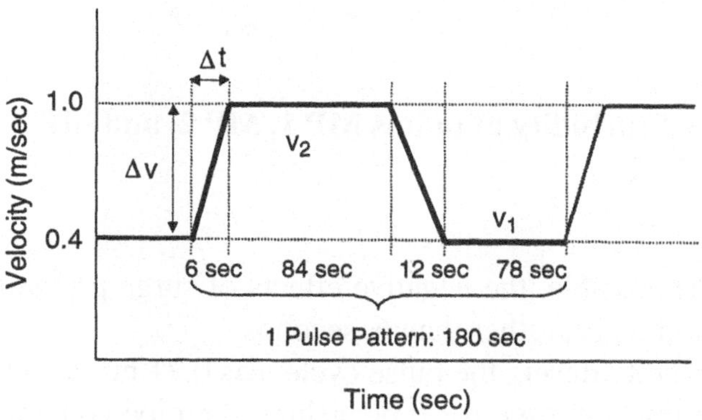

Figure 8.2 Pulse pattern during test

Table 8.1
Combinations of velocity and acceleration that apply for a wall shear stress of $\tau = 3.8$ N/m^2

PVC pipe (outside diameter)	400 mm	160 mm
Internal diameter	376 mm	154 mm
Required constant velocity for $\tau = 3.8$ N/m^2	1.5 m/sec	1.36 m/sec
At basic velocity (v_1) 0.4 m/sec: $\tau =$	0.35 N/m^2	0.42 N/m^2
Required extra τ	3.45 N/m^2	3.38 N/m^2
Related acceleration (a)	0.0365 m/sec^2	0.0868 m/sec^2
$\Delta t = 2$ sec: pulsating velocity (v_2)	0.47 m/sec	0.57 m/sec
$\Delta t = 4$ sec: pulsating velocity (v_2)	0.55 m/sec	0.75 m/sec
$\Delta t = 8$ sec: pulsating velocity (v_2)	0.69 m/sec	1.09 m/sec

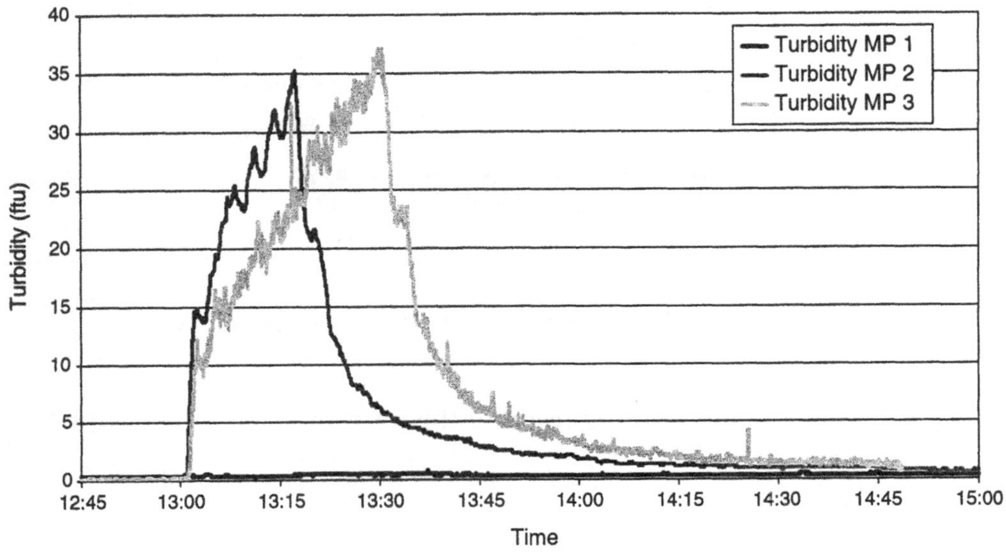

Figure 8.3 Measured turbidity at points MP 1, MP 2, and MP 3 (*Source:* Schaap and van den Boomen 2001)

To avoid, as much as possible, the negative effects of surge pressures, it was decided to shut off the valve over a period of 12, rather than 6, seconds.

The mean velocity during the pulse cycle was 0.71 m/sec. The test section was 1,400 m in length, and thus the flushing time for replenishing the pipe contents three times was 1 hour and 39 minutes. One pulse cycle lasted 3 minutes. Thirty-three pulses were introduced throughout the entire test.

The pressure at MPs 1, 2, and 3 was measured continuously during the test. As a result of opening and closing the regulating valve (throttle valve), pressure surge resulted in a pressure increase of 35 kPa and a pressure decrease of 35 kPa. Turbidity was measured online, at MPs 1, 2, and 3 (see Figure 8.3). Turbidity at MP 1 remained constant during the test period, which confirms that the work was carried out with a clean-water front.

During the measurements, turbidity increased abruptly at the moment when the water was accelerated from 0.4 to 1.0 m/sec for the first time. After the acceleration, which is when the velocity was a constant 1.0 m/sec, turbidity fell immediately, then subsequently increased only after the next acceleration was introduced. Turbidity increased with every pulse to a maximum of approximately 35 ftu at the moment at which the pipe section up to the measuring point was replenished once. Turbidity fell quickly after this, and pulses then had only a minor visible effect on turbidity. At MP 3, turbidity fell after the third replenishment (14 hours, 39 minutes) to approximately 1.4 ftu. The value, then, at MP 2 was approximately 1 ftu. After the test, the pipe was flushed at a velocity of 1.45 m/sec. This showed that a small amount of sediment had remained in the pipe.

Tests with pulsating flushing showed that accelerations during flushing could be introduced in a relatively simple and controllable manner. The sharply increasing turbidity and related turbidity pattern indicate that a clearly observable cleaning effect occurs when a pulse pattern and the associated accelerations are introduced.

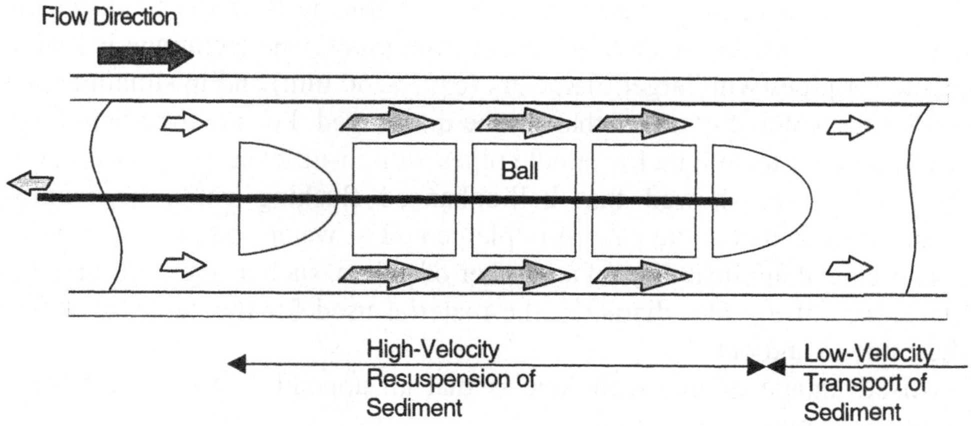

Figure 8.4 Schematic diagram of ball-enhanced flushing

When a pulse pattern of the kind used in the test is introduced, the pressure fluctuations that result from pressure surge are relatively small.

Because the maximum required velocity is 1.0 m/sec, instead of the more common 1.5 m/sec, this method of cleaning can be used in areas where a pipe's carrying capacity is limited. The mean velocity is only 0.7 m/sec, so it is possible to reduce the total volume of flushing water that has to be discharged by 50%.

Further research will be carried out to determine the effect of pulsating flushing on other pipe materials and pipes of different diameters. Moreover, the pulse pattern that is introduced will have to be optimized, so that subsequent flushing is not required and so that a further reduction can be achieved in the maximum velocity (v_2) and in the total volume flow that is required.

Ball-enhanced Flushing

A mean velocity of 1.5 m/sec is required to loosen and resuspend sediment when flushing with water. In the case of ball-enhanced flushing, a local increase in velocity is produced by locally reducing the wet area in the pipe. The reduction is achieved by introducing a series of balls during flushing (see Figure 8.4). The diameter of the balls is chosen so that the velocity at the location of the balls is greater than 1.5 m/sec. Once the sediment has been resuspended, it is carried along with the flow of water, for which a lower velocity is required.

The term *ball-enhanced flushing* was introduced because the first experiments were conducted using balls. In practice, the flushing balls are made of pieces of PE pipe. Balls with a diameter of 250 mm are, for example, used in a 400-mm-diameter PVC pipe. The front and rear ends are closed off using PE sheets. Four holes are made in the rear sheet, so that the flushing balls are able to fill with water. This gives the balls a specific gravity that is practically the same as that of water. The front and rear balls are rounded. A plasticized steel cable joins the balls, and a spacer is fitted between the balls. The balls are connected by a cable to a reel, which is rolled out at a slower rate than the velocity of the water. As ball-enhanced flushing takes place under pressure, a more or less watertight feedthrough is created at the location of the launching fitting. The balls are introduced using a special launching fitting (Y fitting). The balls are then carried along by the flow.

Ball-enhanced flushing was evaluated in various tests in the field, which showed that it could be used to effectively remove sediment from pipes. The technique is mainly considered to be applicable for pipes with larger diameters (e.g., >250 mm) and in situations in which there are limitations on the volume of water that can be discharged. For the time being, pulsating flushing appears to be a better technique for remote pipes with insufficient supply capacity.

The field tests showed that ball-enhanced flushing could be readily implemented. However, before this technique can be implemented at water companies, it will be necessary to improve its practical applicability on a number of points, such as the introduction and removal of the balls in pipes of a wider diameter, the material used for the cable, and the mechanism for rolling the cable in and out.

A disadvantage of this technique is that an appendage (Y fitting) for introducing and removing the balls must be installed.

The maximum length that can be flushed in a single operation is currently 2,400 m. Butterfly valves and changes in diameter impose limitations.

FLOWJET Technique

In the United Kingdom, WRc has developed a technique called FLOWJET, which has been proven to remove films and deposits via a turbulent flushing technique that uses less water than other methods (Marshallsay and Balderson 2002). The system has been used in the field in the United Kingdom and is licensed to AMEC Utilities. The technique, which was developed because of an increasing demand for efficient cleaning systems for water mains, uses a tethered device to block the normal flow in the pipe (Figure 8.5).

The differential head across the device is used to produce high-velocity flow through small apertures in the device. This high-velocity flow is directed at the walls of the pipe by careful alignment of the apertures, and the jets formed are used to clean the pipe. The available pressure in the main and the size of the apertures limit the volume of water that the method uses.

Because the differential pressure across the device produces a force that tries to move the device in the direction of the flow of water in the pipe, it is necessary to restrain the device to prevent it from being carried along with the flow. The tether on the device can then be paid out at the required rate to allow the device to move along the main, cleaning it as it goes.

The debris cleaned from the pipe wall or stirred up from the floor of the pipe will be put into suspension just in front of the device. Any debris or sediment that is light enough to remain in suspension will be carried with the flow in the main in front of the device. Heavier sediment will be continually stirred up and moved forward before settling out again. The device restricts the velocity of the water in the main, meaning that the light deposits will be transported relatively slowly in front of the cleaning head.

The device is used by inserting it through an entry point such as a hydrant tee (with the hydrant removed; Figure 8.6). The tethering cable is led back through a gland to a suitable winch, which is used to hold back the device and stop it from being swept along the main. A downstream hydrant is then opened to perform two functions: (1) reduce the pressure on the downstream side of the device, ensuring that the differential pressure is maximized; and (2) make sure that any debris cleaned from the main is flushed out at this point. To ensure that this is the case, a valve downstream of this hydrant could be closed to make sure that all the flow past the device continues out through the hydrant. The higher the differential head is across the device, the more effective the cleaning action will be.

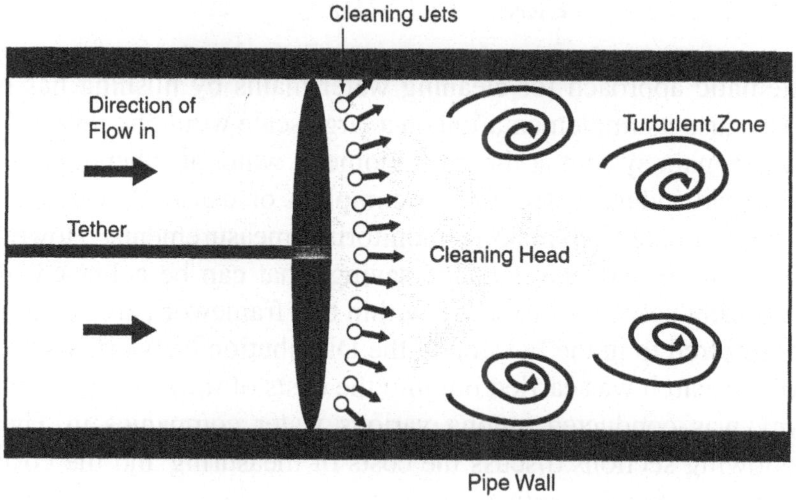

Figure 8.5 Schematic of cleaning head in pipe ***(Source:*** **Marshallsay and Balderson [2002]. Reprinted with permission from WRc plc.)**

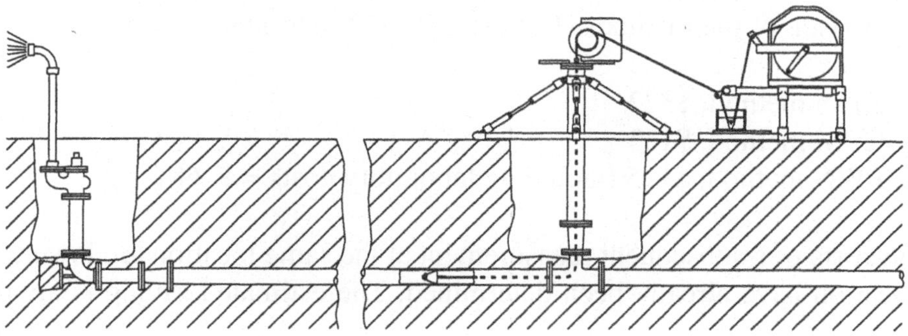

Figure 8.6 Schematic showing operation with an early prototype system ***(Source:*** **Marshallsay and Balderson [2002]. Reprinted with permission from WRc plc.)**

Trials of the FLOWJET System

The system has been extensively tested on flow loops at Swindon in the United Kingdom on a variety of deposits, including sand, grit, loose floc material, chalk, biofilms, and grease. The trials showed the system to be effective at removing loose deposits, sediments, and deposits adhering to the pipe wall.

The system was then tested on a number of live water mains with problems of manganese deposits, iron deposits, biofilms, and chalk deposits. During the field trials, the operational procedures of the system were tested and refined, and the effectiveness of the cleaning operation was modified.

The system in its current form has been developed for cleaning plastic and other smooth pipes in the range of 75 mm to 225 mm in diameter, cleaning lengths of up to 500 m in one pass.

COSTS OF MEASURING VERSUS FLUSHING

A systematic approach for cleaning water mains by flushing has been developed in The Netherlands. Because its implementation on a large scale would be relatively expensive, the aim is to clean the water main system at the right moment, which is when a pipe is fouled to just below the level that would induce customers to complain or cause a reduction in water quality. The moment can be estimated by making monitoring measurements. However, the extra costs of measuring should be in balance with the savings that can be achieved by deferring a cleaning operation. Measuring activities discussed within this framework are concerned with the indicative measurements referred to in the Fouling of the Distribution Network section in chapter 4.

In 2001, research was carried out into the costs of measuring and the costs of flushing, for which a survey was conducted among various water companies in The Netherlands (Schaap 2001). The following sections discuss the costs of measuring and the costs of flushing. The two approaches are then compared with each other.

Costs of Measuring

Three water companies kept records of the time spent and costs incurred in connection with the different measuring methods described in chapter 4. An hourly rate of $37 was assumed for the comparison of the methods. The average costs per method were

- Kiwa method, $322
- OPM method, $133
- PWN method, $115 (including laboratory costs = $168)

The costs of a subsequent measurement at the same location are lower, as the protocol and monitor location have already been determined. The estimated saving is approximately $30 per measurement.

Costs of Flushing

The quantifiable costs of flushing water main systems are determined by, among other factors

- Cleaning method and total time involved
- Costs of a flushing team
- Making measurements and analyses
- Costs of drinking water used for cleaning
- Network modifications, such as fitting flushing points and valves
- Communication costs (to, for example, consumers, the local authority, and the water control authority)
- Possible permits for removing or discharging flushing water, or both

There may also be adverse consequences that are less easy to quantify, such as

- Possible flooding, water damage, and traffic obstructions
- Supply disruption

- Image damage, if consumers see very dirty flushing water coming out of the water main system
- Possible adverse secondary effects of cleaning, such as network contamination caused by inexpert use of fire hydrants or by damage

A survey was conducted among Dutch water companies to obtain an insight into the costs of flushing. The details supplied by seven companies were processed, and the main conclusions were that

- The average cost of cleaning is $550/km, with costs varying between $250/km and $1,050/km.
- Three companies work with a fixed flushing frequency that varies from once every 3 years to once every 6 years.
- The average production rate for cleaning is 2.8 km/day per flushing team, with a variation between 1.4 and 4.5 km/day per flushing team.
- The costs can be divided into an average of 32% for preparation, 62% for performing the work, and 7% for evaluating and measuring effectiveness.
- The highest measuring density for determining the degree of network fouling is one measurement per 2 km, whereas the lowest was one measurement per 27 km.

If an average flushing frequency of once every 4 years is assumed to apply, the annual costs of cleaning come to around $140 per km. When compared with the recorded costs of seven tests carried out in the field under Kiwa's supervision, these costs came to around $800–$1,000/km for flushing for the first time. The attributed costs were considerably higher because extra costs are incurred for drawing up a flushing plan, fitting flushing points, and inspecting fire hydrants and valves. The subsequent flushing operations cost $160–$400/km. The latter roughly corresponds with the average costs derived from the survey, although they are a little higher because the costs of tests in the field are normally higher than the costs of work that is performed regularly.

Number of Measurements

The various companies employed a measuring density that varied from one measurement per 2 km to one per 27 km for measuring the degree of fouling. We can assume that the optimal measuring density lies within this range. To check this, a hypothesis was formulated—that a measuring density of one measurement per 10–20 km of pipe provides sufficient information about the degree of network fouling in a supply area. The hypothesis was considered from two angles of approach.

Approach 1: Reduce Flushing Frequency From 4 to 5 Years. To indicate how much financial leeway is available for making measurements in a supply area, the following example determines the amount that can realistically be saved by reducing the flushing frequency.

- Supply area: 1,000 km
- Current flushing frequency: once every 4 years
- Possible flushing frequency: once every 5 years
- Cleaning costs: $550 per km per operation
- Costs of monitoring measurements: $133 per measurement

Reducing the flushing frequency from once every 4 years to once every 5 years reduces the annual costs of cleaning by $27,500. If this amount were to be spent entirely on monitoring measurements, it would cover the costs of 207 measurements. This would correspond with a measuring density of one measurement per 5 km of pipe.

Approach 2: Determine Acceptable Monitoring Costs per Meter of Pipe. A clear difference of opinion will probably emerge between managers and researchers on the acceptable costs of obtaining information about the degree of network fouling. Some of these individuals from the participating Dutch water company sector were asked what amount they would consider acceptable for performing measurements. This showed that an average of $10/km of pipe was considered acceptable.

Applying $10/km of pipe as the acceptable measuring results in a budget of $10,000/year. This should correspond with a measuring density of one measurement per 13 km of pipe.

The above examples lead to the cautious conclusion that the hypothesis of a measuring density of one measurement per 10–20 km appears to be justified at first sight, and that this would, in turn, lead to an acceptable cost level. The savings achieved by making a realistic reduction in the cleaning frequency appears to be larger than the extra costs of making the measurements. If, after a number of years, a deeper insight is obtained into the degree of fouling in distribution networks, the measuring density can be adjusted.

CHAPTER 9
CONCLUSIONS AND RECOMMENDATIONS

CONCLUSIONS

It is clear that many water utilities deal with discolored water problems. Research into this topic is currently concentrated in the United Kingdom, the United States, and The Netherlands. Despite the large amount of research into several aspects of discolored water, it is still one of the main concerns for water utilities.

Chemical, physical, biological, and hydraulic processes, both in treatment and distribution, all influence the generation of discolored water. Dominant processes can vary from situation to situation. Almost any change in treatment or distribution will have an effect, whether gradual or momentary, on the risk of discolored water.

Processes involved in the generation of discolored water are almost always related. Discolored water can occur only if a source of particulate matter is present and the hydraulic conditions promote sedimentation and occasional resuspension.

The occurrence of discolored water can be monitored on the basis of

- Customer complaints
- Sediment and turbidity monitoring (various sampling techniques)
- Resuspension potential (Kiwa Water Research, The Netherlands)
- Filter paper color method (WRc, United Kingdom)

However, details of each method's relationship to customer perception thresholds are not available.

Customer complaints are very helpful in determining the degree of fouling of the water main system. Using complaints as a measure of fouling requires a proper system for registering and tracking complaints. Such a system includes the following components:

- Uniform and complete registration of the correct information
- Interpretation that allows complaints to be segmented and indicators to be calculated
- Use of complaints for beginning an investigation, testing company standards, and controlling the company's operational processes

In The Netherlands, a great deal of experience has been accumulated with the resuspension potential method. Using this method, it is easy to determine whether resuspended sediment is present in the water main system. Much research has been conducted into the method's optimization, and the method can currently be carried out relatively easily in a short period of time. Refer to Table 4.8 for the resuspension potential criteria that indicate that a water main system is fouled.

GENERAL RECOMMENDATIONS FOR FURTHER RESEARCH

Although the processes involved in the generation of discolored water are known in a general sense, the mechanisms that govern these processes are not fully understood. This is especially true for

- The iron release mechanisms
- The formation of particles within the distribution system and the interrelations among the physical, chemical, and biological processes that determine particle formation
- The role of microbiologically influenced corrosion in the network
- Interactions between corrosion and the biological stability of the water

Because understanding which processes are dominant in which circumstances is key to dealing with discolored water problems, a tool that demonstrates the dominant processes under various circumstances would be useful.

Techniques for Discolored Water Risk Assessment

The resuspension potential method has proven to be a tool that gives information on the fouling of the network. If the shape of the turbidity peak can give useful information on the composition of the sediment in the network, the resuspension potential method must be considered.

In addition, a flowchart that indicates which tests should be done and which data should be analyzed in cases of discolored water problems would be welcome. Work has started in The Netherlands on the development of such a program, but it is not yet available.

Monitoring Water Quality

Because the techniques used in the filtrate scan do not yield any biological information, techniques for monitoring the biostability of the water must be considered.

An integral strategy is needed to simultaneously deal with as many quality-threatening processes as possible. A cost-benefit analysis should determine which control measures should be applied under what circumstances.

Sediment Behavior

The study conclusions on the behavior of sediment in pipes are based on a limited number of measurements, meaning that they are relatively "soft." Further research will pay particular attention to mathematical modeling of sediment transport and will also include the aspect of acceleration.

Research is also recommended into the forces exerted on cohesive sediment in the distribution mains by shear stress. This research should finally result in the development of a model for sediment transport in drinking water pipes. This model could form the basis for developing models of water main system fouling and calculation models of water main systems for designing self-cleaning water main systems.

Cleaning Pipes

Further research is necessary into the effect of pulsating flushing on other pipe materials and diameters. The pulse pattern that must be introduced will also be optimized to make subsequent flushing unnecessary and to achieve a further reduction in the maximum velocity and the total volume flow required.

Tests in the field showed that ball-enhanced flushing could be readily implemented. But before this technique can be implemented at water companies, it will be necessary to improve its practical applicability on a number of points, such as the introduction and removal of the balls in pipes of a wider diameter, the material used for the cable, and the mechanism for rolling the cable in and out.

APPENDIX A
PROTOCOL

1. Prepare for resuspension potential measurement.	
1.1	Describe the pipe: • Diameter, length, material, type • Origin of water • Place and number of valves and fire hydrants
1.2	Determine the length of the pipe section that has to be isolated.
1.3	Determine the required volume flow to obtain a velocity increase of 0.35 m/sec: • in 63-mm diameter: 4 m^3/hr • in 100-mm diameter: 10 m^3/hr • in 150-mm diameter: 23 m^3/hr • in 200-mm diameter: 40 m^3/hr
1.4	Select the fire hydrant at which a disturbance will be introduced. Select the valves that have to be adjusted to isolate the section of pipe.
1.5	Determine the locations at which monitoring systems must be installed. • Monitor location requirements: – A single connection to the water pipe through a tap-in point with a service tap – A single connection to a fire hydrant – Connection to 220 V via a lamppost or domestic supply – A discharge possibility for approximately 100 L of water per hour • Fire hydrant requirements: – Fire hydrant with restrictions
2. Introduce standardized disturbance.	
2.1	Isolate the pipe section by shutting selected valve(s).
2.2	Install the standpipe fitted with the volume flowmeter, ensuring that the standpipe's valve is closed. Do not flush the fire hydrant clean or do so very carefully.
2.3	Maintain the selected volume flow for 5 minutes or 15 minutes. Digitally register: • Volume flow and pressure • Time of opening • Time of closing
2.4	After measuring, open all closed valves again. Record the time when the valves are opened.

APPENDIX B
STATISTICAL COMPARISON OF MEASURING METHODS

Figure B.1 shows the results of the four measuring methods.

In some cases, it may be necessary to translate the results of a measuring method into those of another method, on the basis of a model that describes the relationship between the results of the two measuring methods. To work out these models, we first produced a distribution diagram for each combination of the two methods. This allowed us to visually determine which model is most suitable for describing the relationship between the results of the two measuring methods. The model was then worked out for each combination of two measuring methods by means of regression analysis. We used a robust form of regression analysis for this, the Passing–Bablok method (Passing and Bablok 1983, 1988). This is also suitable for situations in which (1) both of the variables involved display-measuring errors, (2) the measuring errors are not from a normal probability distribution, (3) the standard deviation of the measuring error is not constant over the entire measurement range for one of the two variables concerned, and (4) peaks in the measured values may occur.* Another important advantage of the Passing–Bablok method is that the model that is worked out can be used in two directions.

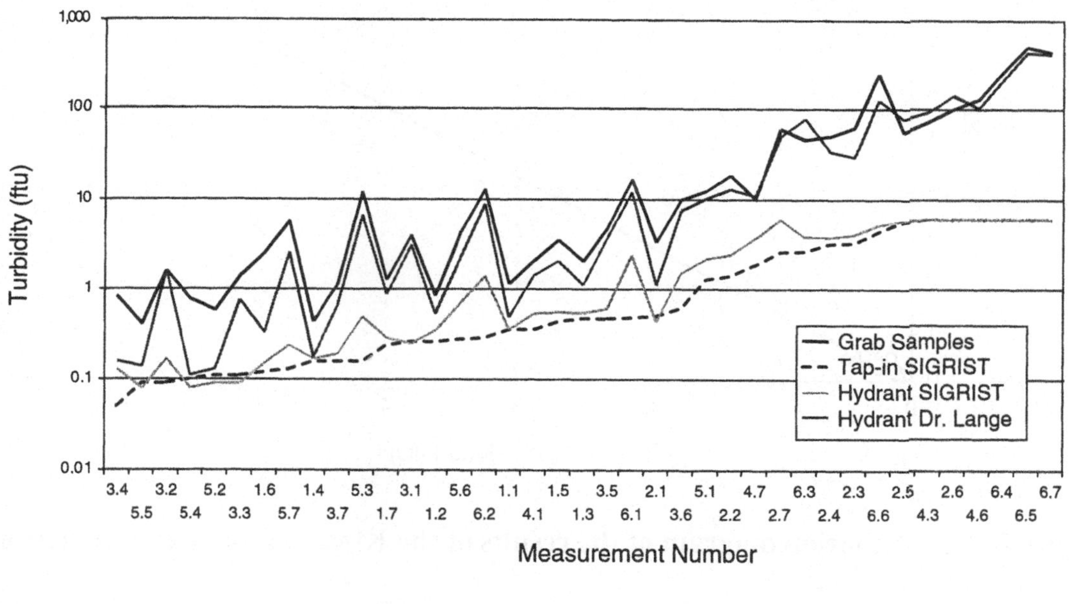

Figure B.1 Graph of the results of the four measuring methods, with the results of Kiwal method arranged in ascending order

* The normal regression analysis is not suitable for any of these situations and produces more or less unusable results.

The six models that were developed were

$$Kiwa2 = -0.0386 + 1.4286 \cdot Kiwa1$$

$$\log_{10}(PWN) = 1.1081 + 1.1316 \cdot \log_{10}(Kiwa1)$$

$$\log_{10}(OPM) = 0.9256 + 1.583 \cdot \log_{10}(Kiwa1)$$

$$\log_{10}(PWN) = 0.9138 + 1.1861 \cdot \log_{10}(Kiwa2)$$

$$\log_{10}(OPM) = 0.6429 + 1.4600 \cdot \log_{10}(Kiwa2)$$

$$\log_{10}(OPM) = -0.3524 + 1.1651 \cdot \log_{10}(PWN)$$

Figures B.2 through B.7 show the distribution diagrams for all six combinations of two measuring methods. The figures also illustrate the changes in the regression model that were worked out for the relationship between the results of the two measuring methods.

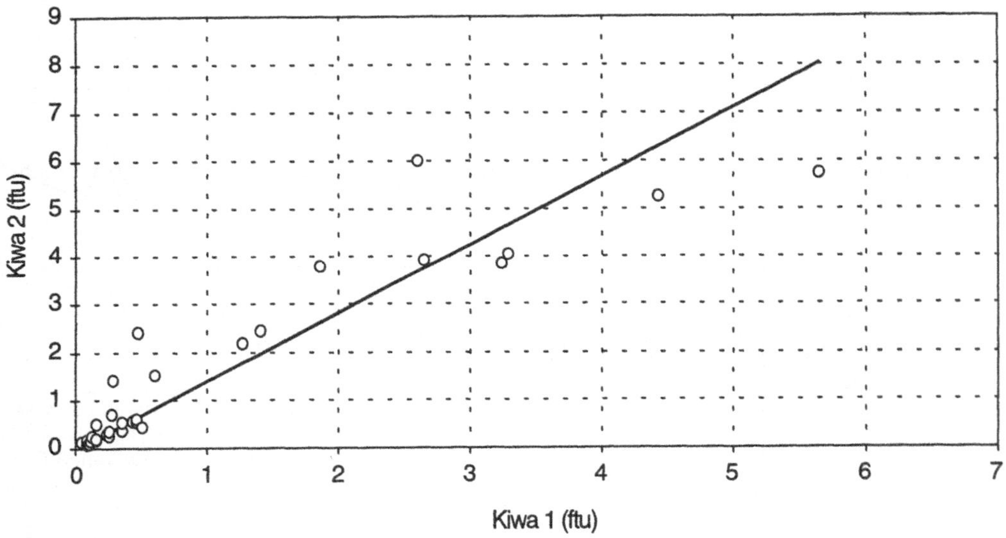

Figure B.2 Distribution diagram of the results of the Kiwa1 method and the Kiwa2 method

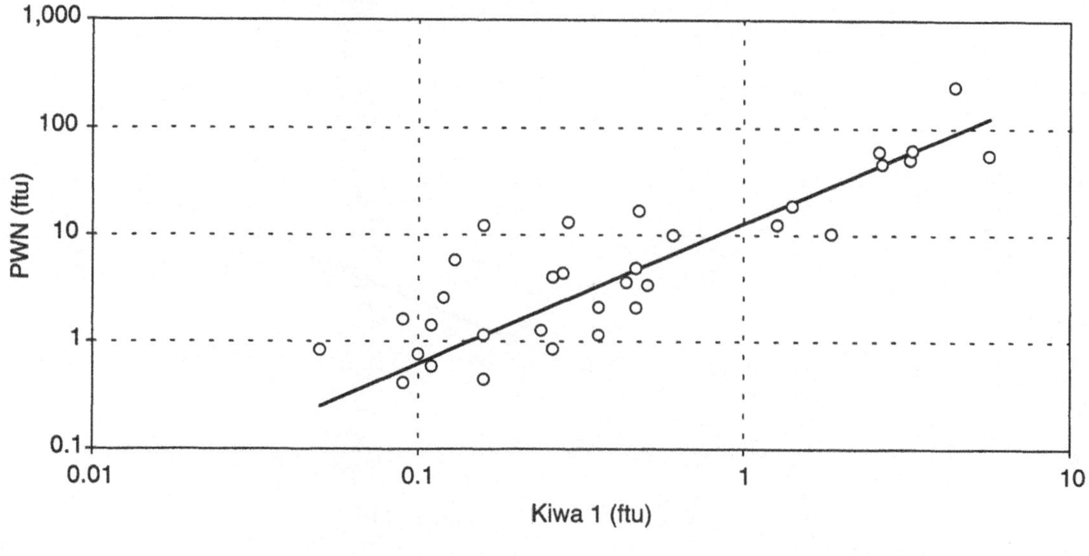

Figure B.3 Distribution diagram of the results of the Kiwa1 method and the PWN method

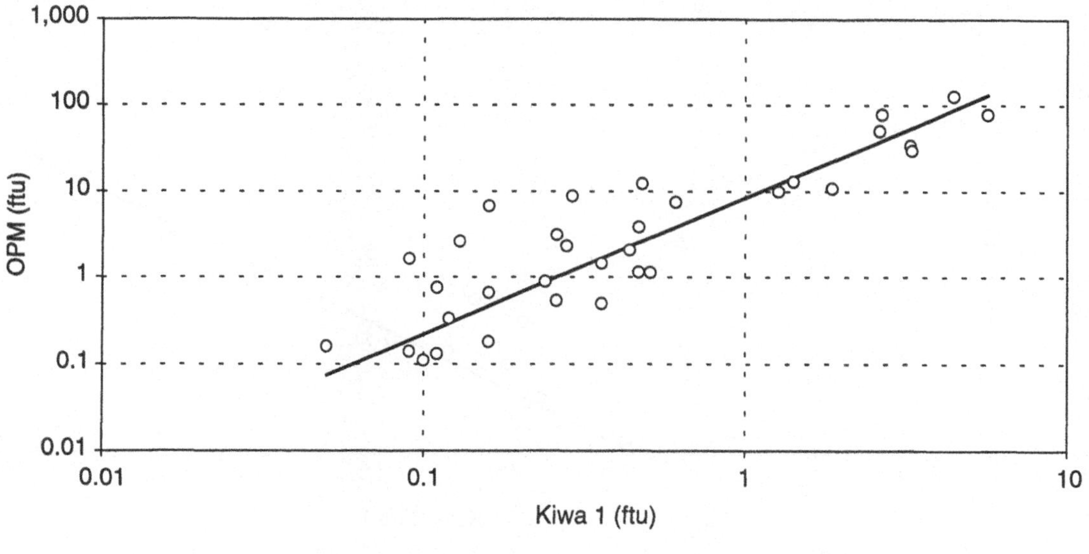

Figure B.4 Distribution diagram of the results of the Kiwa1 method and the OPM method

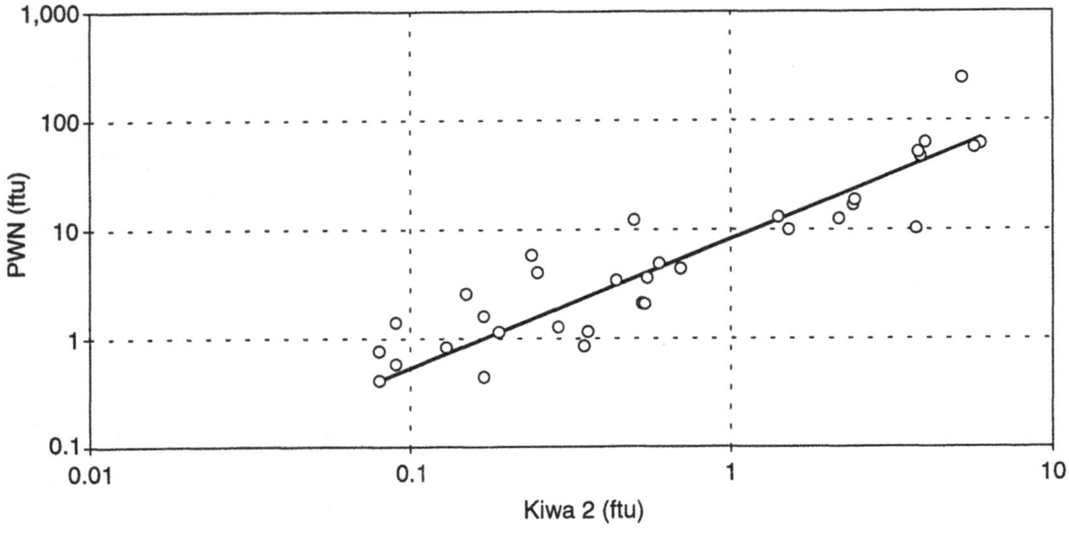

Figure B.5 Distribution diagram of the results of the Kiwa2 method and the PWN method

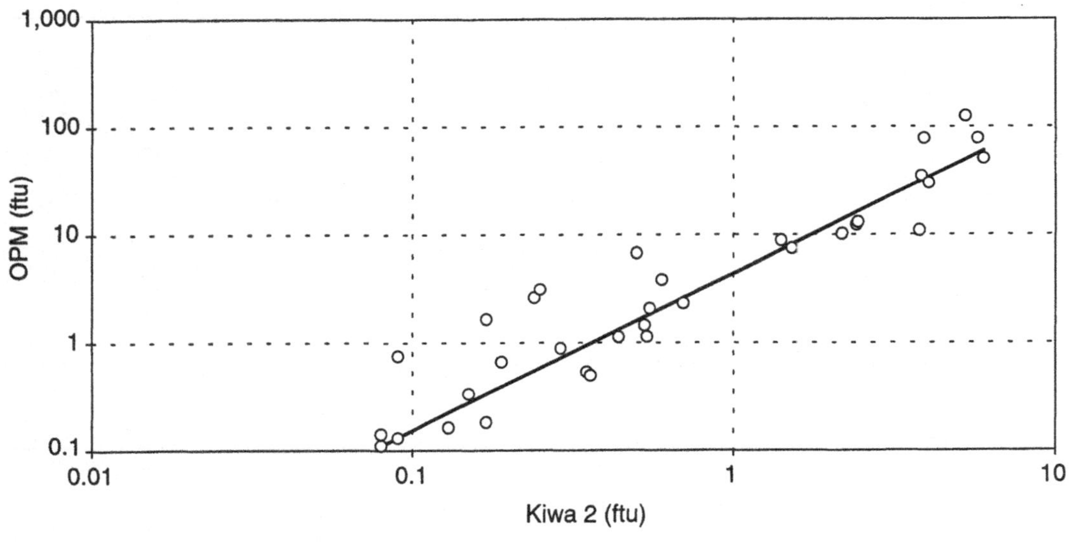

Figure B.6 Distribution diagram of the results of the Kiwa2 method and the OPM method

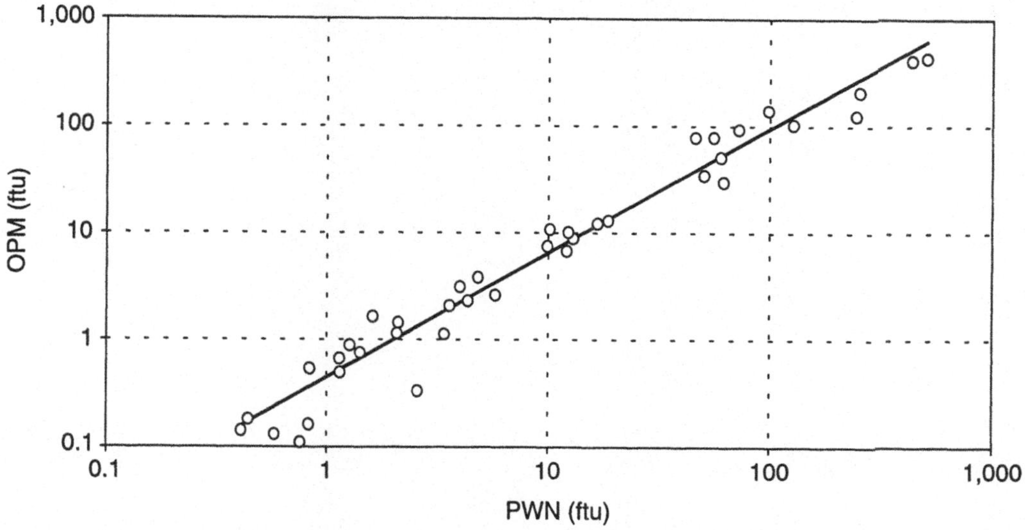

Figure B.7 Distribution diagram of the results of the PWN method and the OPM method

REFERENCES

Antoun, E.N., J.E. Dyksen, and D.J. Hiltebrand. 1999. Unidirectional Flushing—A Powerful Tool. *Jour. AWWA*, 91(7):62–71.

Benjamin, M.M., H. Sontheimer, and P. Leroy. 1996. Corrosion of Iron and Steel. In *Internal Corrosion of Water Distribution Systems*, 2nd ed. Denver, Colo.: AWWA.

Beuken, R.H.S. 2001. *Validatie Nieuwe Ontwerprichtlijnen Voor Distributienetten: Laborato Riumproeven Voor Standaardsediment.* Kiwa report BTO 2001.171 (c). Nieuwegein, The Netherlands: Kiwa Water Research (in Dutch).

Beuken, R.H.S., and J.H.M. van Lieverloo. 2001. *Gedrag van Sediment in Distributieleidingen: Literatuuronderzoek.* Kiwa report BTO 2001.002 (c). Nieuwegein, The Netherlands: Kiwa Water Research (in Dutch).

Boireau, A., P. Mousty, G. Proucelle, and B. Mercier. 2000. Diagnostic of the Water Distribution Network of Perpignan Following the Red Water Phenomenon. *TSM*, 95(11):37–41 (in French).

Building Resolution. 1991. The Hague, The Netherlands: The Ministry of Housing, Spatial Planning and the Environment.

Clement, J., M. Hayes, P. Sarin, W.M. Kriven, J. Bebee, K. Jim, M. Beckett, V.L. Snoeyink, G.J. Kirmeyer, and G. Pierson. 2002. *Development of Red Water Control Strategies.* Denver, Colo.: AwwaRF.

Cordonnier, J., H. Madiec, O. Ziegler, P. Soukathchoff, and P. Leroy. 2000. Diagnosis of Old Cast Iron Network and Protection of the Internal Walls With Water Treatment. *TSM*, 95(11):41–47 (in French).

Cury, A. 1994. National Report Brazil. *Water Supply*, 12(1):2. IR 7-9, IR 7-12.

Davis, J.P., and J.C. Smith. 1993. Implementing a Programme to Manage Distribution Water Quality. In *Proc. of the AWWA Water Quality Technology Conference.* Part I, 217–231. Denver, Colo.: AWWA.

Edwards, M., S. Jacobs, and R.J. Taylor. 2000. The Blue Water Phenomenon. *Jour. AWWA*, 92(7):71–82.

Gauthier, V., C. Rosin, L. Mathieu, J. Portal, J. Block, P. Chaix, and D. Gatel. 1996. Characterization of the Loose Deposits in Drinking Water Distribution Systems. In *Proc. of the AWWA Water Quality Technology Conference*. Denver, Colo.: AWWA.

Gray, N.F. 1994. *Drinking Water Quality: Problems and Solutions*. Chichester, U.K.: Wiley.

Gregory, J., and D.W. Nelson. 1984. A New Optical Method for Flocculation Monitoring. In *Solid–Liquid Separation.* Chichester (West Sussex): Published for the Society of Chemical Industry, London, by E. Horwood, New York, NY.

Guerber, F., J. Magning, and A. Larigauderie. 2000. Operation of a Water Distribution Network in a Rural Zone—Application to the Case of Corrosive Scaling Water. *TSM*, 95(11):32–36 (in French).

Hambsch, B., and I. Wagner. 1998. Water Quality in Distribution. *National Report Germany Water Supply,* 16(1/2):89–130.

Hedberg, T. 1994. National Report Sweden. *Water Supply*, 12(1):2. IR 7-36, IR 7-38.

Kirmeyer, G.J., M. Friedman, J. Clement, A. Sandvig, P.F. Noran, K.D. Martel, D. Smith, M. LeChevallier, C. Volk, E. Antoun, D. Hiltebrand, J. Dyksen, and R. Cushing. 2000. *Guidance Manual for Maintaining Distribution System Water Quality.* Denver, Colo.: AwwaRF.

Kreukniet, A.A. 2001. *Vergelijking Meetwaarden Verkorte Meetmethoden.* Kiwa report BTO 2001.199 (c). Nieuwegein, The Netherlands: Kiwa Water Research (in Dutch).

Kuch, A., H. Sontheimer, and I. Wagner. 1983. Die Messung der Aufeisenung im Trinkwasser—Ein neues Verfahren zur beurteilung vom Deckschichten in Schwarzen Stahlrohren. *Werkst. u. Korros.*, 34:107.

Liebeschuetz, J., and A.D. Hulsmann. 1990. Special Problems in Rehabilitation, Quantifying Discoloured Water Problems and the Assessment of Air Scouring as a Remedial Action. *Water Supply,* 8(Berlin):181–186.

Lytle, D.A., and V.L. Snoeyink. 2000. The Effect of Water Chemistry on the Physical Properties of Iron Colloids and Aqueous Iron Suspensions. In *Proc. of the International Distribution Research Symposium.* Denver, Colo.: AWWA.

Marshallsay, D., and P. Balderson. 2002. *FLOWJET*—A New Technology to Solve an Old Problem. U.K. patent 2300242.

McNeill, L.S., and M. Edwards. 2000. Phosphate Waters and Red Water in Stagnant Iron Pipes. In *Proc. of the International Distribution Research Symposium.* Denver, Colo.: AwwaRF.

McNeill, L.S., and M.E. Edwards. 2001. Iron Pipe Corrosion in Distribution Systems. *Jour. AWWA,* 93(7):88–100.

Passing, H., and W. Bablok. 1983. A New Biometrical Procedure for Testing the Equality of Measurements From Two Different Analytical Methods. *J. Clin. Chem. Clin. Biochem.,* 1983(21):709–720.

Passing, H., and W. Bablok. 1988. A General Regression Procedure for Method Transformation. *J. Clin. Chem. Clin. Biochem.*, 1988(26):783–790.

Rhodes Trussell, R., and I. Wagner. 1996. Corrosion of Galvanized Pipe. In *Internal Corrosion of Water Distribution Systems,* 2nd ed. Denver, Colo.: AWWA.

Rosenthal, L.P.M., D. van der Kooij, and J.H.G. Vreeburg. 2000. The Trends and Prospects of In-pipe Water Quality Control Technologies. Paper presented at the 5th International Symposium on Water Supply Technology, Japan Water Research Center, Kobe, Japan.

Sander, A. 1996. Iron Corrosion in Drinking Water Distribution Systems—A Corrosion Rate Study. Master's thesis. Chalmers University of Technology, Göteborg, Sweden.

Schaap, P.G. 2001. *Kosten van Meten—Resultaten Enquête Schoonmaakkosten.* Kiwa report 2001.208(c). Nieuwegein, The Netherlands: Kiwa Water Research (in Dutch).

Schaap, P.G., and G.H. Ekkers. 1999. *SchoonmaakbEleid Eindhoven, Bepaling Omvang Schoonmaakprogramma.* Kiwa report KOA 99.004. Nieuwegein, The Netherlands: Kiwa Water Research (in Dutch).

Schaap, P.G., and M. van den Boomen. 2001. *Pulserend Spuien Hydron Zuid-Holland.* Kiwa report BTO2001.153(c). Nieuwegein, The Netherlands: Kiwa Water Research (in Dutch).

Schumacher, P.G., I. Wagner, and V. Wehle. 1993. Zentrale Dosierung von Inhibitoren zum Trinkwasser unter ökologischen und technischen Aspecten am Beispiel der Stadtwerke Götingen AG. *GWF-Wasser/Abwasser,* 134(10):628–635.

Shields, A. 1936. *Anwendung der Senlichtkeitsmechanik un der Turbulezforschung auf die Geschiebebewegung.* Berlin, Germany: Mitteilungen der Preussichen Verscuhsanstalt fur Wasserbau und Schiffbau.

Siegers, W., and G. IJpelaar. 2002. *Verbetering Pompstations, Verlaging Spuikosten.* Kiwa report BTO 2001.200 (c). Nieuwegein, The Netherlands: Kiwa Water Research (in Dutch).

Slaats, P.G.G. 2000. *Comparison of Turbidity Meters for Quick Measurement of Resuspension Potential.* Kiwa report SWI 00.122. Nieuwegein, The Netherlands: Kiwa Water Research (in Dutch).

Smith, S.E., A. Bisset, J.S. Colbourne, D. Holt, and B.J. Lloyd. 1997. The Occurrence and Significance of Particles and Deposits in a Drinking Water Distribution System. *Jour. NEWWA*, 111:135–150.

Smith, S.E., D. Holt, T. Ta, and A. Bisset. 2000. Assessing and Controlling Discoloration. In *Proc. of the International Distribution Research Symposium*. Denver, Colo.: AwwaRF.

Smith, S.E., T. Ta, D.M. Holt, A. Delanoue, J.S. Colbourne, A.H.L. Chamberlain, and B.J. Lloyd. 1999. A Pipeline Testing Facility for the Examination of Pipe-Wall Deposits and Red Water Events in Drinking Water. *J. CIWEM*, 1999(13):Feb. 7–15.

Taylor, R.J., R.J. O'Halloran, and P.H. Cannington. 1996. The Causes and Treatment of Blue Water in Copper Plumbing Systems. In *Proc. of the 13th Int'l. Corros. Congr.* Paper 144/1–7. Melbourne, Australia: Australasian Corrosion Association.

van den Boomen, M., and A. van Mazijk. 2002. Schone Leidingen Door Dynamiek van Afnamepatronen. *H_2O*, 5:28–30 (in Dutch).

van den Boomen, M., and J.H.G. Vreeburg. 1999. *Nieuwe Ontwerprichtlijnen Voor Distributienetten.* Kiwa report SWE 99.011, Kiwa Water Research, Nieuwegein, The Netherlands (in Dutch).

van den Hoven, Th.J.J., D. van der Kooij, J.H.G. Vreeburg, and H. Brink. 1994. Methods to Analyse and to Cure Water Quality Problems in Distribution Systems. *Water Supply*, 12(3/4):151.

van den Hoven, Th.J.J., and M.W.M. van Eekeren. 1988: *Optimal Composition of Drinking Water.* Kiwa report 100. Nieuwegein, The Netherlands: Kiwa Water Research.

van der Kooij, D., J.H.M. van Lieverloo, J. Schellart, and P. Hiemstra. 1999. Maintaining Quality Without a Disinfectant Residual. *Jour. AWWA*, 91(1):55–64.

van der Kooij, D., and H.R. Veenendaal. 1993. Assessment of the Biofilm Formation Potential of Synthetic Materials in Contact With Drinking Water During Distribution. In *Proc. of the AWWA Water Quality Technology Conference*. Denver, Colo.: AWWA.

van Lieverloo, J.H.M., D. van der Kooij, C.J.M. Coertjens, and G. Veenendaal. 2002. Sediments in Drinking Water Distribution Systems in The Netherlands. *Jour. AWWA* (forthcoming).

Vreeburg, J.H.G. 1996. *Red Water, Cleaning, Cause and Consequences.* Kiwa report SWE 96.008. Nieuwegein, The Netherlands: Kiwa Water Research (in Dutch).

Vreeburg, J.H.G. 2000. Mechanisms and Solutions for Red Water. In *Proc. of the International Distribution Research Symposium*. Denver, Colo.: AwwaRF.

Wagner, I., and A. Kuch. 1984. The Influence of Water Parameters on Corrosion Rate, Scale Deposition, and Iron-(III) Uptake in Unprotected Iron Pipes. *Water Supply*, 2(3/4):SS 11-1.

Water Quality Statistics. 2000. Rijswijk, The Netherlands: VEWIN (Netherlands Water Companies Association).

Williams, S.M. 1990. Water Quality Problems—The Use of Sodium Silicate and Sodium Polyphosphate to Control Water Quality Problems. *Water Supply*, 8(Berlin):195–198.

Woodward, A., T. Ta, J. Colbourne, and D. Holt. 1995. Behaviour of Particles in a Large Scale Experimental Pipe System. In *Proc. of the AWWA Water Quality Technology Conference.* Denver, Colo.: AWWA.

ABBREVIATIONS

A–C	asbestos cement
ALFI	automatic liquid filtration instrument
ATP	adenosine triphosphate
AWWA	American Water Works Association
AwwaRF	Awwa Research Foundation
BFP	biofilm formation potential
BFR	biofilm formation rate
BTO-Research	joint research program of water companies in The Netherlands
DVGW	German Technical and Scientific Association on Gas and Water
EBI	Engler Bunte Institute
FPC	filter paper color
ftu	formazine turbidity unit (comparable with ntu)
HMN	Hydron Central Netherlands
IR	infrared
ISO	International Organization for Standardization
LED	light-emitting diode
M	measurement of fouling the network
MP	measuring point
NOM	natural organic matter
NRE-WATER	Nutsbedrijven Regio Eindhoven
ntu	nephelometric turbidity unit
OPM	simplified resuspension potential method developed with BTO-Research group
PE	polyethylene
Pt–Co	platinum–cobalt (unit of color)
PVC	polyvinyl chloride
PWN	Provincial Water Company of North Holland
RPM	resuspension potential measurement

TIC	total inorganic carbon
TOC	total organic carbon
TU	tap unit
TZW	Technology Zentrum Wasser
VEWIN	Netherlands Water Companies Association
WGld	Water Company Gelderland
WMD	Water Company Drenthe
WML	Water Company Limburg
WMO	Water Company Overijssel
WOB	Water Company East Brabant

IWA Publishing's authorised EU representative for General Product Safety Regulations is Diane D'Arras, 15 rue Duret, 75116 Paris, France, e-mail: safety@iwap.co.uk.

Printed and bound by CPI Group (UK) Ltd, Croydon, CR0 4YY
23/09/2026
02221720-0004